Systems Biology in Animal Production and Health, Vol. 2

Haja N. Kadarmideen
Editor

Systems Biology in Animal Production and Health, Vol. 2

Springer

Editor
Haja N. Kadarmideen
Faculty of Health and Medical Sciences
University of Copenhagen
Frederiksberg C, Denmark

ISBN 978-3-319-43330-1 ISBN 978-3-319-43332-5 (eBook)
DOI 10.1007/978-3-319-43332-5

Library of Congress Control Number: 2016956674

Printed on acid-free paper

This Springer imprint is published by Springer Nature
The registered company is Springer International Publishing AG Switzerland
The registered company address is Gewerbestrasse 11, 6330 Cham, Switzerland

Foreword

The increased prominence of "systems biology" in biological research over the past two decades is arguably a reaction to the reductionist approach exemplified by the genome sequencing phase of the Human Genome Project. A simplistic view of the genome projects was that the genome sequence of a species, whether humans, model organisms, plants or farmed animals, represents a blueprint for the organism of interest, and thus characterising the sequence would reveal the relevant instructions. Subsequent targets for the reductionist or cataloguing approach were complete lists of transcripts (transcriptomes) and proteins (proteomes) for the organism of interest. The 'omics approach to the comprehensive characterisation of an organism, tissue or cell has also been extended to metabolites and hence metabolomes. A catalogue of parts, however, is insufficient to understand how an organism functions. Thus, a holistic approach that recognises the interactions between components of the system was required. Given the size and complexity of the data and the possible interactions, it was necessary to use advanced mathematical and computational methods to attempt to make sense of the data. Thus, "systems biology" in the 'omics era is widely considered to concern the use of mathematical modelling and analysis together with 'omics data (genome sequence, transcriptomes, proteomes, metabolomes) to understand complex biological systems. The predictive aspect of these models is viewed as particularly important. Moreover, it is desirable that the models' predictions can be tested experimentally. Systems biology, therefore, contributes in part to converting large 'omics data sets from data-driven biology experiments into testable hypotheses.

Systems approaches and the use of predictive mathematical models in biological systems long pre-date the post genome project (re-)emergence of systems biology. Population biologists/geneticists, epidemiologists, agricultural scientists, quantitative geneticists and plant and animal breeders have been developing and successfully exploiting predictive mathematical models and systems approaches for decades.

Quantitative geneticists and animal breeders, for example, have been remarkably successful at developing statistical animal models that are effective predictors of future performance. For decades, these successes were achieved without any knowledge of the underlying molecular components. The accuracy of these models has been increased by using high-density molecular (single nucleotide polymorphism, SNP) genotypes in so-called genomic selection. However, whilst the sequences and genome locations of the SNP markers are known little is known about the functional impact or relevance of the individual SNP loci. Further improvements could be achieved through the use of genome sequence data and by adding knowledge of the likely effects of the sequence variants whether coding or regulatory. Thus, there is a growing commonality between the systems approaches of quantitative geneticists and animal breeders and the 'omics version of systems biology.

Animals are not only complex biological systems but also function within wider complex systems. The recognition that an animal's phenotype is determined by a combination of its genotype and environmental factors simply restates the latter. The environmental factors include, amongst others, feed, pathogens and the microbiomes present in the gastrointestinal tract and other locations. The 'omics technologies allow not only the characterisation of the components of the animal of interest, but also those of its commensal microbes and the microbes, including pathogens present in its environment.

As noted earlier, it is desirable that the mathematical models developed in systems biology are predictive and that the associated hypotheses are testable. Genome editing technologies which have been demonstrated in farmed animal species facilitate hypothesis testing at the level of modifying the genome sequence that determines components of the system of interest.

This volume of *Systems Biology in Animal Production and Health*, edited by professor Haja Kadarmideen, explores some aspects of both quantitative genetics and 'omics led approaches to applying systems approaches to tackling the challenges of improving animal productivity and reducing the burden of disease. The book contains some chapters with R codes and other computer programs, workflow/pipeline for processing and analysing multi-omic datasets from lab all the way to interpretation of results. Hence, this book would be useful particularly for students, teachers and practitioners of integrative genomics, bioinformatics and systems biology in animal and veterinary sciences.

Villa-Vialaneix et al. (chapter "Depicting Gene Co-expression Networks Underlying eQTL") address the challenge of identifying the gene networks that capture the interaction between genes from eQTL data. The application of systems approaches to specific traits of interest in agriculture and biology are reviewed by *Schroyen et al.* (chapter "Applications of Systems Biology to Improve Pig Health"), *Fukumasu et al.* (chapter "Systems Biology Application in Feed Efficiency in Beef Cattle"), and *Vailati-Riboni et al.* (chapter "Nutritional Systems Biology to Elucidate Adaptations in Lactation Physiology of Dairy Cows"). The analysis of transcriptomic data and specifically RNA-Seq data are described in greater detail by *Mazzoni* and *Kadarmideen* (chapter "Computational Methods for Quality Check, Preprocessing and Normalization of RNA-Seq Data for Systems Biology and Analysis").

Finally, farmed animal species are not only important for agriculture but are also used for basic biological research and as models in biomedical research. *Mashayekhi et al.* (chapter "Systems Biology and Stem Cell Pluripotency: Revisiting the Discovery of Pluripotent Stem Cell") describe a systems perspective on pluripotency.

Professor Alan L. Archibald FRSE
Deputy Director, Head of Genetics and Genomics
The Roslin Institute and Royal (Dick) School of Veterinary Studies
University of Edinburgh
Easter Bush, Midlothian EH25 9RG, UK

Preface

Systems biology is a research discipline at the crossroad of statistical, computational, quantitative and molecular biology methods. It involves joint modeling, combined analysis and interpretation of high-throughput omics (HTO) data collected at many "levels or layers" of the biological systems within and across individuals in the population. The systems biology approach is often aimed at studying associations and interactions between different "layers or levels", but not necessarily one layer or level in isolation. For instance, it involves study of multidimensional associations or interaction among DNA polymorphisms, gene expression levels, proteins or metabolite abundances. With modern HTO biotechnologies and their decreasing costs, hugely comprehensive multi-omic data at all "levels or layers" of the biological system are now available. This "big data" at lower costs, along with development of genome scale models, network approaches and computational power, have spearheaded the progress of the systems biology era, including applications in human biology and medicine. Systems biology is an established independent discipline in humans and increasingly so in animals, plants and microbial research. However, joint modeling and analyses of multilayer HTO data, in large volumes on a scale that has never been seen before, has enormous challenges from both computational and statistical points of view. Systems biology tackles such joint modeling and analyses of multiple HTO datasets using a combination of statistical, computational, quantitative and molecular biology methods and bioinformatics tools. As I wrote in my review article (*Livestock Science* 2014, 166:232–248), systems biology is not only about multilayer HTO data collection from populations of individuals and subsequent analyses and interpretations; it is also about a philosophy and a hypothesis-driven predictive modeling approach that feeds into new experimental designs, analyses and interpretations. In fact, systems biology revolves and iterates between these "wet" and "dry" approaches to converge on coherent understanding of the whole biological system behind a disease or phenotype and provide a complete blueprint of functions that leads to a phenotype or a complex disease.

It is equally important to introduce, alongside systems biology, the sub-discipline of *systems genetics* as a branch of systems biology. It is akin to considering "genetics" as a sub-discipline of "biology". It is well known that quantitative genetics/genomics links genome-wide genetic variation with variation in disease risks or a performance (phenotype or trait) that we can easily measure or observe in a population of individuals. However, systems genetics or systems genomics not only performs such genome-wide association studies (GWAS), but also performs linking genetic variations (e.g. SNPs, CNVs, QTLs etc.) at the DNA sequence level with variation in molecular profiles or traits (e.g. gene expression or metabolomic or proteomic levels etc. in tissues and biological fluids) that we can measure using high-throughput next- and third-generation biotechnologies. The systems genetics approach is still "genetics", because we are looking at those genetic variants that exert their effects from DNA to phenotypic expression or disease manifestations through a number of intermediate molecular profiles. Hence, systems genetics derives its name, as originally proposed in my earlier article (*Mammalian Genome*, 2006, 17:548–564), by being able to integrate analyses of all underlying genetic factors acting at different biological levels, namely, QTL, eQTL, mQTL, pQTL and so on. I have provided a complete up-to-date review and illustration of systems genetics or systems genomics and multi-omic data integration and analyses in our review paper published in *Genetics Selection Evolution* (2016), 48:38. Overall, systems genetics/genomics leads us to provide a holistic view on complex trait heredity at different biological layers or levels.

Whether it is systems biology or systems genetics, the gene ontology annotation is one of the most important and valuable means of assigning functional information using standardized vocabulary. This would include annotation of genetic variants falling into functional groups such as trait QTL, eQTL, mQTL, pQTL. Molecular pathway profiling, signal transduction and gene set enrichment analyses along with various types of annotations form the "icing on cake". For this purpose, several bioinformatics tools are frequently used. Most chapters in this book and its associated volume cover these aspects.

I would like to point out that systems biology approaches have been proven to be very powerful and shown to produce accurate and replicable discoveries of genes, proteins and metabolites and their networks that are involved in complex diseases or traits. In very practical terms, it delivers biomarkers, drug targets, vaccine targets, target transcripts or metabolites, genetic markers, pathway targets etc. to diagnose and treat diseases better or improve traits or characteristics in animals, plants and humans. In the world of genomic prediction and genomic selection, there have been an increasing number of studies that have shown high accuracy and predictive power when models include functional QTLs such as eQTL, mQTL, pQTL which, in fact, are results from systems genetics methods.

This book and its associated volume cover the above-mentioned principles, theory and application of systems biology and systems genetics in livestock and animal models and provides a comprehensive overview of open source and commercially available software tools, computer programing codes and other reading materials to learn, use and successfully apply systems biology and systems genetics in animals.

Overall, I believe this book is an extremely valuable source for students interested in learning the basics and could form as a textbook in higher educational institutes and universities around the world. Equally, the book chapters are very relevant and useful for scientists interested in learning and applying advanced HTO studies, integrative HTO data analyses (e.g. eQTLs and mQTLs) and computational systems biology techniques to animal production, health and welfare. One of the chapters focuses on stem cell research in animal models elucidating systems biology of pluripotency with translational applications for human neurological and brain diseases. The two volumes of this book is a result of contributions from highly reputed scientists and practitioners who originate from renowned universities and multinational companies in the UK, Denmark, France, Italy, Australia, USA, Brazil and India. I would like to thank the publisher Springer for inviting me to edit two volumes on this subject, publishing in an excellent form and promoting the book across the globe. I am grateful to all contributing authors and co-authors of this book. I also wish to thank Ms. Gilda Kischinovsky from my research group for proofreading and the staff at Springer involved in production of this book. Last but not least, I wish to thank my wife and children who have given me moral support and strength while I reviewed and edited this book.

Copenhagen, Denmark — Haja N. Kadarmideen
September 2016

Contents

Depicting Gene Co-expression Networks Underlying eQTLs

Nathalie Villa-Vialaneix, Laurence Liaubet, and Magali SanCristobal

Abstract

Deciphering the biological mechanisms underlying a list of genes whose expression is under partial genetic control (i.e., having at least one eQTL) may not be as easy as for a list of differential genes. Indeed, no specific phenotype (e.g., health or production phenotype) is linked to the list of transcripts under study. There is a need to find a coherent biological interpretation of a list of genes under (partial) genetic control. We propose a pipeline using appropriate statistical tools to build a co-expression network from the list of genes, then to finely depict the network structure. Graphical models are relevant because they are based on partial correlations, closely linked with causal dependencies. Highly connected genes (hubs) and genes that are important for the global structure of the network (genes with high betweenness) are often biologically meaningful. Extracting modules of genes that are highly connected permits a significant enrichment in one biological function for each module, thus linking statistical results with biological significance. This approach has been previously used on a pig eQTL dataset (Villa-Vialaneix et al. 2013) and was proven to be highly relevant. Throughout the present chapter, we define statistical notions linked with network theory, and apply them on a reduced dataset of genes with eQTL that were found in the pig species to illustrate the basics of network inference and mining.

N. Villa-Vialaneix (✉)
MIAT, Université de Toulouse, INRA, Castanet Tolosan, France
e-mail: nathalie.villa@toulouse.inra.fr

L. Liaubet • M. SanCristobal
GenPhySE, Université de Toulouse, INRA, INPT, INP-ENVT, Castanet Tolosan, France
e-mail: laurence.liaubet@toulouse.inra.fr; magali.san-cristobal@toulouse.inra.fr

H.N. Kadarmideen (ed.), *Systems Biology in Animal Production and Health, Vol. 2*,
DOI 10.1007/978-3-319-43332-5_1

1 Introduction

In the search for genetic mechanisms underlying production or health phenotypes (e.g., terminal), GWAS studies have been intensively used, and have shown their limits. Classical tools in integrative biology aim at discovering links between terminal phenotypes and fine phenotypes (e.g., transcriptome, proteome, metabolome), in huge numbers. Integrating both approaches is possible: searching for a genetic basis of fine phenotypes (e.g., eQTL, mQTL studies). The step further goes back to the terminal phenotypes with the precious and fine knowledge acquired with omics data. The focus of this chapter is linked to integrative biology and eQTL studies. The common pipeline for differential analysis is the use of linear models for testing differential expression at each gene, followed by a correction for multiple testing. This provides a list of genes whose expressions vary with the phenotype of interest. Then, a functional analysis is performed: GO terms and KEGG pathways; in addition, bibliographic mining is also interesting. The major limitation of this is the incomplete annotation encountered in livestock species: there may be only a part of transcripts that could not be given a gene name (e.g., 78 % in our pig transcripts have a gene name and about half have an associated function), mandatory for bibliographic mining.

eQTL studies provide genetic markers (the so-called eQTLs) that have partial control of gene expression, and a list of genes whose expression is partially under genetic control (genes with eQTL). Upstream, there is some genetic control; genetic markers (the eQTLs) are often observed displayed in genomic clusters (e.g., (Liaubet et al. 2011)). Downstream, a transcriptional control exists followed by a regulation of biological functions. Focusing on genes whose expression is genetically controlled (at least partially), we would like to address some questions. Do they also cluster? Is there a link between clusters of co-expression and biological functions?

The most appropriate tool to achieve this goal is networks. Given the strong loss of information with bibliographic networks (incomplete annotation), an alternative is co-expression networks. Indeed, this statistical approach is based on all expression information, independent of the annotation. There exists various kinds of co-expression networks. We will see in the following that graphical Gaussian models (GGM, based on partial correlation) are very appropriate, in the sense that they are close to causative biological meaning.

After inferring the network in a sparse manner, it is of high interest to mine its structure. Extracting interesting genes (e.g., highly connected, with high incidence on the global structure) can give clues for further biological hypotheses and future experiments. Extracting modules can lead to an enrichment in biological functions, making the link between statistical results and biological interpretation. The functional annotation of the modules, based on a limited number of genes (because of the poor annotation), can then give insights into possible biological functions for unnannotated genes ("guilt by association" approach, see (Dozmorov et al. 2011) and (Gillis and Pavlidis 2012) for a study which questions this approach).

In the article (Villa-Vialaneix et al. 2013), the pipeline briefly described above highlighted key genes, and showed a strong enrichment of one biological function per module. Moreover, one module was linked with meat pH, a particularly interesting phenotype, since it is related to meat production and quality. In this chapter, we will present in detail the overall approach, explaining key aspects linked with network analysis, applying them on a subset of genes with eQTLs extracted from the one studied in (Villa-Vialaneix et al. 2013).

This chapter is organized as follows: Sect. 2 provides basic definitions and concepts for network studies. Section 3 deals with network inference and Sect. 4 deals with network mining. Finally, Sect. 5 deals with biological interpretation of the results. Throughout this article, a small example study is performed using the free statistical software R: codes and datasets are available at http://nathalievilla.org/bio_network.

2 Basic Definitions and Concepts for Graphs/Networks

2.1 Networks

A *network*, also frequently called a *graph*, is a mathematical object used to model *relationships* between *entities*. In its simplest form, it is composed of two sets (V, E):

- The set $V = \{v_1, \ldots, v_p\}$ is a set of p nodes, also called *vertices* that represent the entities.
- The set E is a subset of the set of node pairs, $E \subset \{(v_i, v_j),\ i, j = 1, \ldots, p,\ i \neq j\}$: the node pairs in E are called *edges* of the graph and model a given type of relationships between two entities.

In the following, nodes will be genes and edges will represent a relationship (e.g., co-expression) between two genes. A network is often displayed as in Fig. 1: the nodes are represented with circles and the edges with straight lines connecting two nodes.

This lesson's scope is restricted to simple networks, i.e., to undirected graphs (the edges do not have any direction), with no loop (there is no edge between a given node and itself) and simple edges (there is one edge at most between a pair of nodes). But networks can deal with many other types of real-life situations:

- *Directed graphs* in which the edges have a direction, i.e., the edge from the node *vi* to the node *vj* is not the same as the edge from the node *vj* to the node *vi*. In this case, the edges are often called *arcs*.
- *Weighted graphs* in which a weight (often positive) is associated to each edge.
- *Graphs with multiple edges* in which a pair of nodes can be linked by several edges that can eventually have different labels or weights to model different types of relationships.

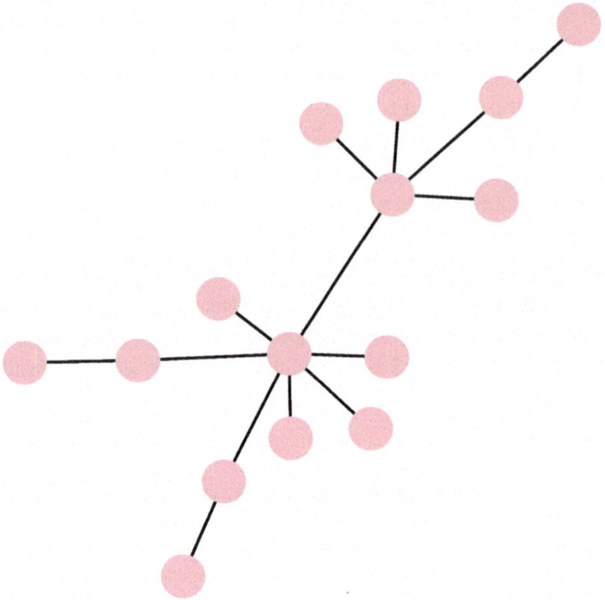

Fig. 1 Example of the representation of a simple network with 15 nodes and 13 edges

- *Labeled graphs* (or graph with node attributes) in which one or several labels are associated to each node, labels can be factors (e.g., a gene function) or numeric values (e.g., gene expression).

2.2 Overview of Standard Issues for Network Analysis

This chapter will address two main issues posed by network analysis:

- The first one will be discussed in Sect. 3 and is called *network inference*: giving data (i.e., variables observed for several subjects or objects), how to build a network whose edges represent the "direct links" between the variables? The nodes in the inferred network are the genes and the edges represent a strong "direct link" between the two gene expressions.
- The second issue comes when the network is already built or directly given: the practitioner then wants to understand the main characteristics of the network and to extract its most important nodes, groups, etc. This ensemble of methods, studied in Sect. 4, is called *network mining* and comprises (among other problems):
  - *Network visualization*: when displaying a network, no a priori position is associated with its nodes and the network can thus be displayed in many different ways.

- *Node clustering*: an intuitive way to understand a network structure is to focus not on individual connections between nodes but on connections between densely connected groups of nodes. These groups are often called *clusters* or *communities* or *modules* and many works in the literature have focused on the problem of extracting these clusters.

2.3 eQTL Data

Throughout this chapter, a subset of genes analyzed in (Villa-Vialaneix et al. 2013) will be used to illustrate the basics of network inference and mining. The applications will be performed using the free statistical software environment http://r-project.org R (version 3.2.5). The packages used are:

- huge (version 1.2.7) for network inference
- igraph (version 1.0.1) for creating network objects and for network mining

The reader interested in this topic may also want to have a look at the "gRaphical Models in R" task view,[1] where he/she will find further interesting packages.

To illustrate key steps, we propose the analysis of a small subset of data in (Liaubet et al. 2011; Villa-Vialaneix et al. 2013), which is a subset of 68 genes having at least one eQTL. This data will be refered to as "68-eqtl" throughout the chapter. This dataset can be downloaded at http://nathalievilla.org/doc/csv/subsetEQTL.csv. The dataset consists of gene expressions for a "small" list of genes (transcripts). It is represented by the matrix $\mathbf{X}$:

$$n\,\text{individuals}\left\{\mathbf{X} = \underbrace{\begin{pmatrix} \cdot & \cdot & \cdot & \cdot & \cdot & \cdot \\ \cdot & \cdot & X_i^j & \cdot & \cdot & \cdot \\ \cdot & \cdot & \cdot & \cdot & \cdot & \cdot \end{pmatrix}}_{p \text{ variables (gene expressions)}}\right.,$$

where Xij is the expression quantification of gene j in individual i. Even restricting to a small subset of genes, having $n < p$ is the standard situation which, as discussed later, poses some problems for network inference. These data can be loaded using the following command line:

```
expression = read.csv("data/subsetEQTL.csv", row.names=1)
```

if the dataset provided at http://nathalievilla.org/doc/csv/subsetEQTL.csv is stored in subdirectory "data" of R working directory.

The boxplots of the $p = 68$ variables (genes) of the "68-eqtl" dataset are displayed in Fig. 2 (left). The correlation matrix between the 68 genes is displayed in Fig. 2 (right) showing that a potential structure has to be highlighted.

[1] https://cran.r-project.org/web/views/gR.html.

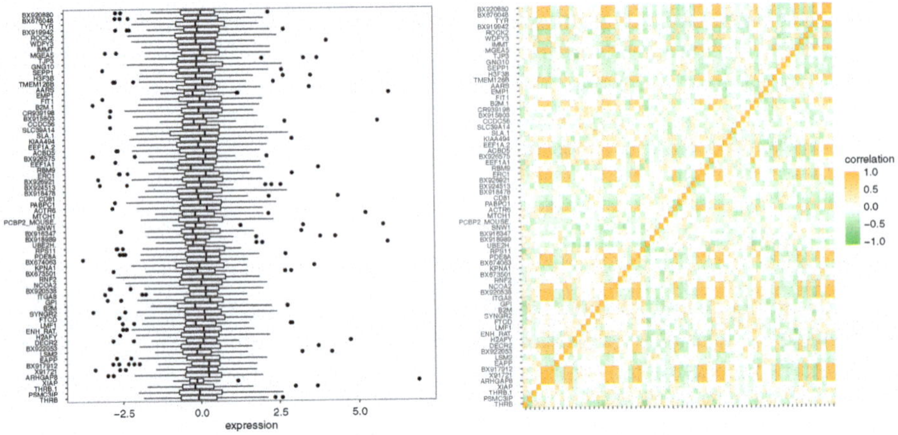

Fig. 2 *Left*: boxplot of the gene expression distributions (68 genes). *Right*: heatmap of the correlation matrix between pairs of gene expressions

3 Network Inference

The aim of this section is to choose an appropriate type of network, then to infer the network based on data (expression of the 68 genes). In short, "inferring a network" means building a graph for which

- The nodes represent the p genes.
- The edges represent a "direct" and "strong" relationship between two genes. This kind of relationships aims at tracking hierarchical influence and possible transcriptional or genetic regulations.

The main advantage of using networks over raw data is that such a model focuses on "strong" links and is thus more robust. Also, inference can be combined/compared with/to bibliographic networks to incorporate prior knowledge into the model but, unlike bibliographic networks, networks inferred from one of the models presented below can handle even unknown (i.e., not annotated) genes into the analysis.

Even if alternative approaches exist, a common way to infer a network from gene expression data is to use the steps described in Fig. 3:

1. First, the user calculates pairwise similarities (correlations, partial correlations, information-based similarities such as the mutual information) between pairs of genes.
2. Second, the smallest (or less significant) similarities are thresholded (using a simple threshold chosen by a given heuristic or a test or sparse approaches with penalization while calculating the similarities or other more sophisticated methods).

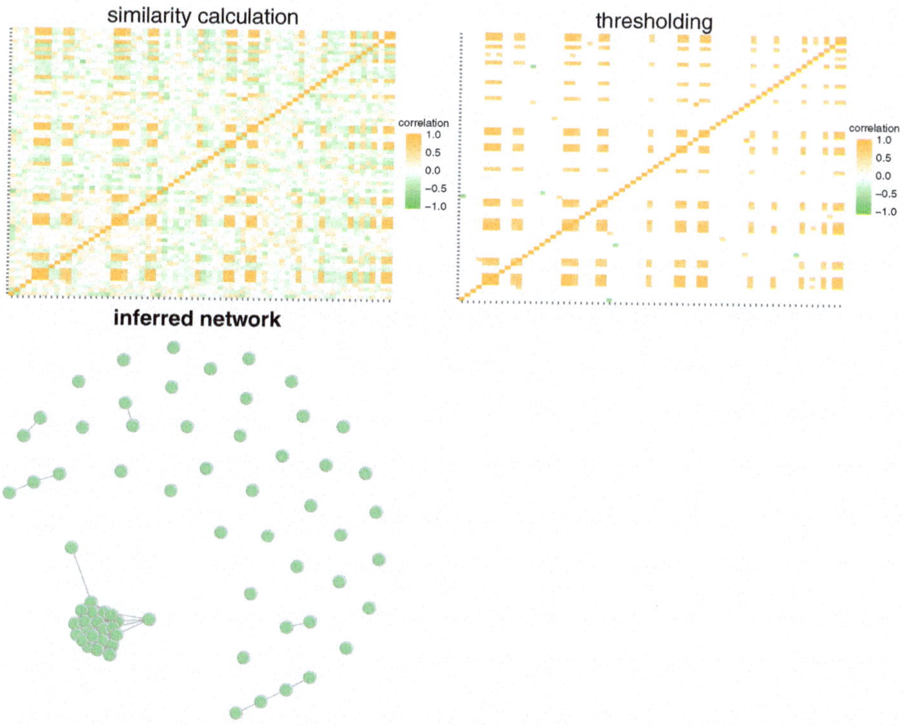

Fig. 3 Main steps in network inference

3. Lastly, the network is built from the non-zero similarities, putting an edge between two genes with a non-zero similarity (which thus correspond to the highest values, in a given sense that depends on the thresholding method, of the similarity).

This approach leads to produce *undirected* networks. Additionaly, the edges of the network can be weighted by the strength of the relationship (i.e., the absolute value of the similarity) and signed by the sign of the relation (i.e., if the similarity is positive or negative). This approach is used in (Kogelman et al. 2015) to integrate DE genes and eQTL genes in a single co-expression network related to obesity in pigs.

3.1 Limits of the Pearson Correlation

A simple, naive approach to infer a network from gene expression data is to calculate pairwise correlations between gene expressions and then to simply threshold the smallest ones, possibly, using a test of significance. This approach is sometimes called *relevance network* (Butte and Kohane 1999, 2000). The R package **huge**[2] can

[2] http://cran.r-project.org/web/packages/huge.

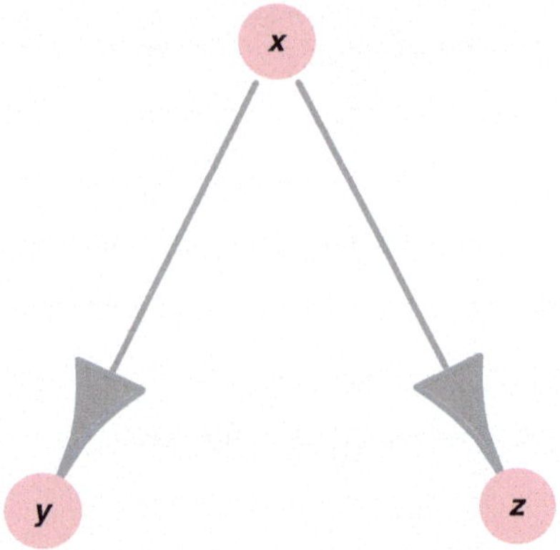

Fig. 4 Small model showing the limit of the correlation coefficient to track regulation links

be used to infer networks in such a way. However, if easy to interpret, this approach may lead to strongly misunderstanding the regulation relationships between genes. To better understand the problem posed by using direct correlations in network inference, we will discuss the simple situation described in Fig. 4. In this model, a single gene, denoted by x, strongly regulates the expression of two other genes, y and z. This situation is well illustrated using the simple mathematical model.

Figure 4 is a small model showing the limit of the correlation coefficient to track regulation links: when two genes y and z are regulated by a common gene x, the correlation coefficient between the expression of y and the expression of z is strong as a consequence. For instance,

$$X \sim \mathcal{U}[0,1], \qquad Y \sim 2X + 1 + \varepsilon_1 \text{ and } Z \sim -2X + 2 + \varepsilon_2$$

in which $\mathcal{U}[0,1]$ is the uniform distribution in [0, 1], and ε_1 and ε_2 are independent and centered Gaussian random variables independent of X with a standard deviation equal to 0.1. A quick simulation with R gives the following results:

```
x = rnorm(100)
y = 2*x+1+rnorm(100,0,0.1)
cor(x,y)

    ## [1] 0.9988261

z = -2*x+1+rnorm(100,0,0.1)
cor(x,z)

    ## [1] -0.998756

cor(y,z)

    ## [1] -0.9980506
```

Hence, even though there is no direct (regulation) link between z and y, these two variables are highly correlated (the correlation coefficient is larger than 0.99) as a result of their common regulation by x.

3.2 Partial Correlation and Gaussian Graphical Model (GGM)

This result is unwanted and using a *partial correlation* can deal with such strong indirect correlation coefficients. The partial correlation between y and z is the correlation between the expression of y and z, *knowing the expression of x*. In the above example, it is equal to the correlation between the residuals of the linear models:

$$Y = \beta_1 X + \varepsilon_1 \text{ and } Z = \beta_2 X + \varepsilon_2$$

and in our case, it is equal to

```
cor(lm(z~x)$residuals,lm(y~x)$residuals)

    ##    [1]    -0.1933699
```

which is much smaller than the direct correlation, while the other two partial correlations remain large:

```
cor(lm(x~y)$residuals,lm(z~y)$residuals)

    ##    [1]    -0.6208908

cor(lm(x~z)$residuals,lm(y~z)$residuals)

    ##    [1]    0.6481373
```

When using partial correlation, the *conditional dependency graph* is thus estimated. Under a Gaussian model (see (Edwards 1995) for further explanations), in which the gene expressions $X = \left(X^j\right)_{j=1,\ldots,p}$ are supposed to be distributed as centered Gaussian random variables with covariance matrix Σ, this graph is defined as follows:

$$v_j \leftrightarrow v_{j'} \left(\text{genes } j \text{ and } j' \text{are linked}\right) \Leftrightarrow \mathrm{Cor}\left(X^j, X^{j'} \mid \left(X^k\right)_{k \neq j,j'}\right) \neq 0$$

in which the last quantity is called *partial correlation*, $\pi_{jj'}$. In this framework, $\mathbf{S} = \Sigma^{-1}$ is called the *concentration matrix* and is related to the partial correlation $\pi_{jj'}$ between Xj and $X^{j'}$ by the following relation:

$$\pi_{jj'} = -\frac{\mathbf{S}_{jj'}}{\sqrt{\mathbf{S}_{jj}\mathbf{S}_{j'j'}}}. \tag{1}$$

This equation indicates that non-zero partial correlations (i.e., edges in the conditional dependency graph) are also non-zero entries of the concentration matrix **S**.

3.3 Estimating the Conditional Dependency Graph with Graphical LASSO

The empirical estimator Σ of Σ is calculated from the $n \times p$ matrix of gene expression $\mathbf{X}$ generated from the Gaussian distribution $\mathcal{N}(0,\Sigma)$,

$$\Sigma_{jj'} := \frac{1}{n}\sum_i \left(\mathbf{X}_i^j - \bar{X}^j\right)^2 \text{ with } \bar{X}^j = \frac{1}{n}\sum_i \mathbf{X}_i^j,$$

calculated from the observations $\mathbf{X}$. A major issue when using Σ^{-1} for estimating $\mathbf{S}$ is that the empirical estimator Σ is ill-conditioned because it is calculated with only a small number n of observations: the sample size n is usually much lower than the number of variables p. Hence, Σ^{-1} is a poor estimate of $\mathbf{S}$ and must not be used as it is.

Several attempts to deal with such a problem have been proposed. The seminal work (Schäfer and Strimmer 2005a, b) uses shrinkage, i.e., $\mathbf{S}$ is estimated by $\hat{\mathbf{S}} = \left(\Sigma + \lambda \mathbb{I}\right)^{-1}$ (for a given small $\lambda \in \mathbb{R}^+$). Then, the obtained partial correlations are thresholded either by choosing a given thresholding value or a given number of edges or by using a test statistics presented in (Schäfer and Strimmer 2005a), which is itself based on a Bayesian model. This method is implemented in the R package **GeneNet**.[3]

The previous method is a two-step method which first estimates the partial correlations and then selects the most significant ones. An alternative method is to simultaneously estimate and select the partial correlations using a sparse penalty. It is known under the name Graphical LASSO (or GLasso). Under a GGM framework, partial correlation is also related to the estimation of the following linear models:

$$X^j = \sum_{k \neq j} \beta_k^j X^k + \varepsilon_j \tag{2}$$

by the relation

$$\beta_k^j = -\frac{\mathbf{S}_{jk}}{\mathbf{S}_{jj}}$$

which, combined with Eq. (1) shows again that non-zero entries of the linear model coefficients correspond exactly to non-zero partial correlations.

Hence, several authors (Friedman et al. 2008; Meinshausen and Bühlmann 2006) have proposed to integrate a sparse penalty in the estimation of (2) by ordinary least squares (OLS):

$$\forall j = 1, \ldots, p, \qquad \arg\min_{\beta^j} \left[\sum_{i=1}^n \left(\mathbf{X}_i^j - \sum_{k \neq j} \beta_k^j \mathbf{X}_i^k \right)^2 + \lambda \left\| \beta^j \right\|_{L^1} \right] \tag{3}$$

[3] https://cran.r-project.org/web/packages/GeneNet.

where $\|\beta^j\|_{L^1} = \sum_{k \neq j} |\beta^j_k|$ is the L_1-norm of $\beta^j \in \mathbb{R}^{p-1}$, which is added to the OLS minimization problem in order to force only a restricted number of non-zero entries in βj. λ is a regularization parameter that controls the sparseness of βj (the larger λ, the fewer the number of non-zero entries in βj). It is generally varied during the learning process and the most adequate value is selected. This method is implemented in the R package **huge**.

Finally, several approaches have been proposed to deal with the choice of a proper λ: (Liu et al. 2010) proposes the StARS approach, which is based on a stability criterion, while (Lysen 2009) and (Foygel and Drton 2010) propose approaches based on a modification of the BIC criterion. All these methods are implemented in the R package **huge**.

3.4 Application

Using the "68-eqtl" data, a network can be inferred using the method described in (Meinshausen and Bühlmann 2006) with the R package **huge**. The package is loaded with

```
library(huge)
```

The concentration matrix is estimated for several values of λ with:

```
glassoRes = huge(as.matrix(expression), nlambda=100,
    method="glasso")
```

The option `nlambda` is used to set the number of regularization parameter values λ used for the estimation. The result is a list of estimated concentration matrices (one for each value of λ, whose sparsity decreases when λ decreases), stored in `glassoRes$icov`. These matrices are (almost) all sparse, which means that most of their entries are equal to zero (the matrices obtained with small λ contains much fewer zeros than the ones with larger λ).

To select one of the 100 concentration matrices, the function `huge.select` implements several model selection methods. Among them, the "StARS" method chooses the largest λ so that the obtained concentration matrix is replicable with random subsampling. More precisely, many random subsamples are generated and a criterion is computed to assess the stability of any given edges in the inference obtained from all subsamples. The most sparse graph which is still stable according to these criteria is the one chosen by the method. This approach can be used with:

```
glassoFinal = huge.select(glassoRes, criterion="stars")
```

which results in an object that contains the optimal value of lambda, `glassoFinal$opt.lambda` (here equals to 0.3551), the optimal 68×68

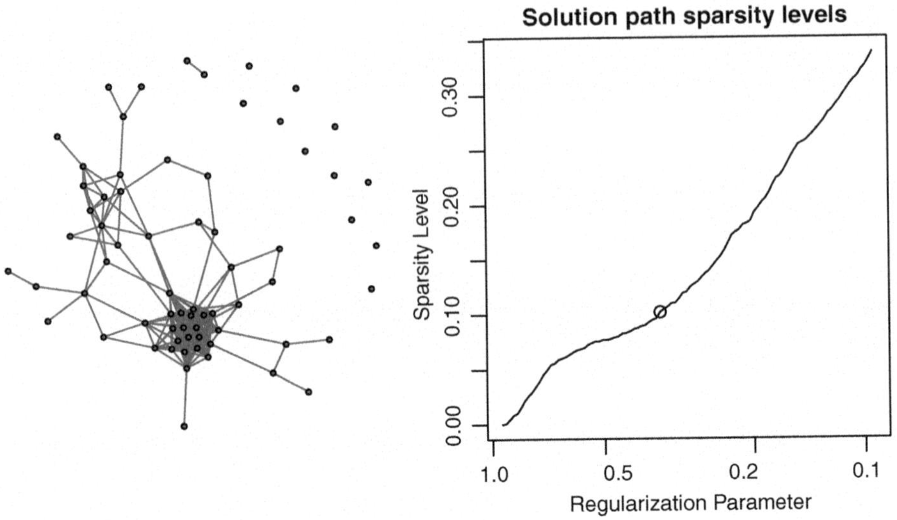

Fig. 5 Summary of the result of the "StARS" selection method. *Left*: selected network. *Right*: solution sparsity (% of inferred edges over the number of pairs of nodes in the graph) versus λ. The chosen λ is emphasized with a dot on the curve

concentration matrix in `glassoFinal$opt.icov` and the optimal sparse adjacency matrix of the inferred network in `glassoFinal$refit`. The result of the selection is summarized in Fig. 5, which is produced by the following command line:

```
plot(glassoFinal)
```

Finally, a network R object can be obtained for further studies using the R package **igraph**. More precisely, the function graph_from_adjacency_matrix can be used on the sparse adjacency matrix glassoFinal$refit and the function simplify is used to remove multiple edges and loops.

```
glassoNet = graph_from_adjacency_matrix(glassoFinal$refit, mode="max")

glassoNet = simplify(glassoNet)
glassoNet

## IGRAPH U--- 68 232 -
## + edges:
## [1]  1--18  1--27  1--31  1--40  1--41  2--17  4--8   4--11  4--62  5--6
## [11] 5--7   5--11  5--19  5--20  5--21  5--26  5--39  5--40  5--43  5--44
## [21] 5--52  5--56  5--63  5--64  5--65  5--67  5--68  6--7   6--10  6--11
## [31] 6--19  6--20  6--25  6--26  6--39  6--40  6--43  6--44  6--56  6--61
## [41] 6--67  6--68  7--10  7--11  7--19  7--20  7--21  7--26  7--34  7--35
## [51] 7--39  7--40  7--43  7--44  7--46  7--52  7--56  7--61  7--63  7--65
## [61] 7--67  7--68  9--29  10--11 10--21 10--25 10--34 10--39 10--43 10--44
## [71] 10--49 10--61 10--67 10--68 11--19 11--20 11--21 11--25 11--34 11--35
## [81] 11--39 11--40 11--43 11--44 11--67 11--68 12--28 12--46 12--64 13--18

## + ... omitted several edges
```

This graph (an igraph object) contains $p = 68$ nodes and 232 edges.

Gene names (included in the column names of the expression matrix) can be attached to the nodes as an attribute called "name" which is then easily used when displaying the network or selecting nodes. This setting is performed with the function V:

```
V(glassoNet)$name = colnames(expression)
```

As shown in Fig. 5, the inferred network is composed of several groups of nodes that are not connected with each other. These groups are called the *connected components of the graph*. Using **igraph**, they can be extracted with the function components:

```
glassoComp = components(glassoNet)
head(glassoComp$membership)
```

```
##   THRB   PSMC3IP   THRB.1   XIAP   ARHGAP8   X91721
##     1        1        2       1        1         1
```

```
glassoComp$csize
```

```
##  [1]  55  1  2  1  1  1  1  1  1  1  1  1  1
```

```
glassoComp$no
```

```
##  [1]  13
```

The inferred network has glassoComp$no=13 connected components, most of them composed of only one node. The largest connected component has glassoComp$csize=55 nodes. The number of the connected component of a given gene in the gene network is given in glassoComp$membership and the connected components can thus be obtained with the function induced_subgraph:

```
glassoSubNet = induced_subgraph(glassoNet,
    glassoComp$membership==which.max(glassoComp$csize))
```

Finally, the largest connected component of the inferred network, which contains 55 nodes and 231 edges, will be named "55-eqtl network" in the sequel. This network is the one that will be studied further in the next section which is devoted to network mining. This graph can be exported into an external format, such as the widely used "graphml" format, with the function write_graph

```
write_graph(glassoSubNet, file="results/lcc.graphml",
            format="graphml")
```

The obtained file can then be imported in most softwares dedicated to graph mining for exploratory purposes. More information about the possible formats for graph exportation is available with

```
help(write_graph)
```

4 Network Mining

In this section, a graph $\mathcal{G} = (V, E)$ is supposed to be given, where $V = \{v_1, \ldots, v_p\}$ is the set of nodes and E is the set of edges. Mining a network is the process in which the user extracts information about the most important nodes or about groups of nodes that are densely connected.

4.1 Network Visualization

Visualization tools are used to display the graph in a meaningful and aesthetic way. Standard approaches in this area use *force directed placement* (FDP) algorithms (see (Fruchterman and Reingold 1991), among others). The principle of these algorithms can be illustrated by an analogy to the following physical mechanism which:

- Attaches attractive forces to the edges of the graph (similar to springs) in order to force connected nodes to be represented close to each other.
- Attaches repulsive forces between all pairs of nodes (similar to electric forces) to force nodes to be displayed separately.

The algorithm performs iteratively from an (usually random) initial position of the nodes until stabilization. The R package **igraph** (see (Csardi and Nepusz 2006)) implements several layouts and even several FDP based layouts for static representation of the network.

Using **igraph**, the network inferred in Sect. 3 can be displayed using the functions `layout.fruchterman.reingold` (for calculating the layout with the FDP method of (Fruchterman and Reingold 1991)) and `plot.igraph` (for displaying it on a graphical device). The result of the function `layout.fruchterman.reingold` is a matrix with two columns and 55 rows that contains the positions of the nodes. It can be attached to the `igraph` object as a graph attribute named "layout" to be used when passed to the function `plot` (Fig. 6). Several characteristics of the graph representation, that are related to nodes and edges (colours, shapes, labels…), can be defined in the `plot.igraph` options.

```
glassoSubNet$layout =
    layout.fruchterman.reingold(glassoSubNet)
plot(glassoSubNet, vertex.size=0,
     vertex.label.color="black",
     vertex.label.cex=0.8)
```

More information on the plot.igraph options are provided in the help:

```
help(igraph.plotting)
```

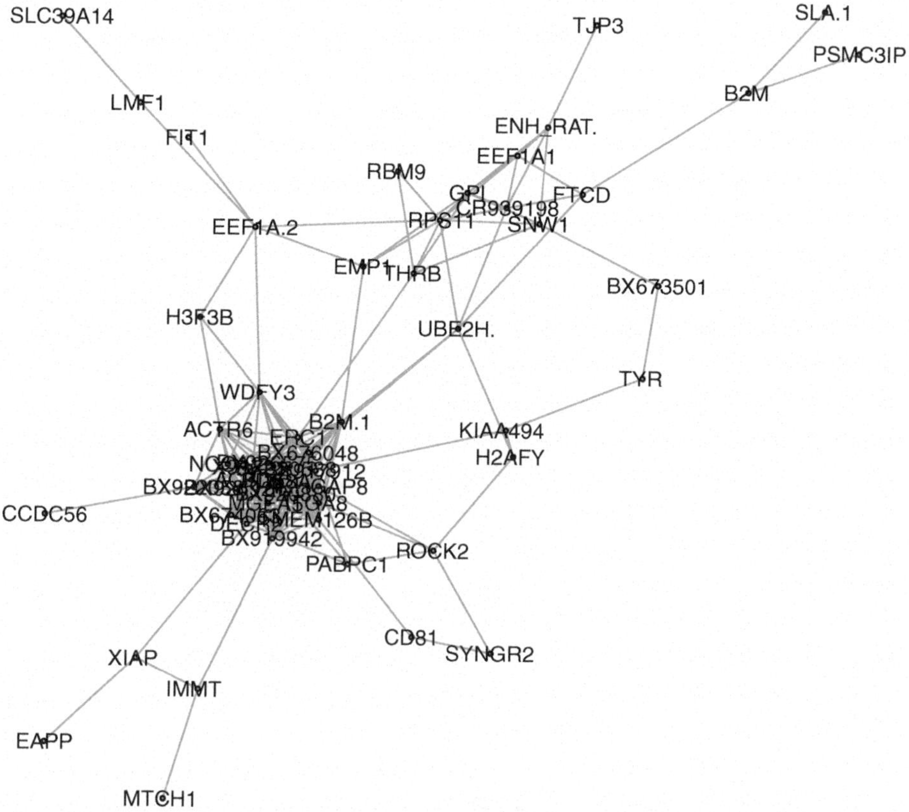

Fig. 6 Representation of the inferred network with Fruchterman and Reingold force directed placement algorithm

The free softwares Gephi[4] (Bastian et al. 2009), Tulip[5] (Auber 2003) or Cytoscape[6] (Shannon et al. 2003), among others, can also be used to visualize a network interactively (they support zooming and panning, among other features).

4.2 Global Characteristics

This section gives the definition of two global numerical characteristics that can help to understand the network structure.

Definition 1 (density) The *density* of a network is the number of edges divided by the number of pairs of nodes, $\frac{|E|}{p(p-1)/2}$.

[4] http://gephi.org.

[5] http://tulip.labri.fr.

[6] http://www.cytoscape.org.

Fig. 7 Simple network with a transitivity equal to 1/3

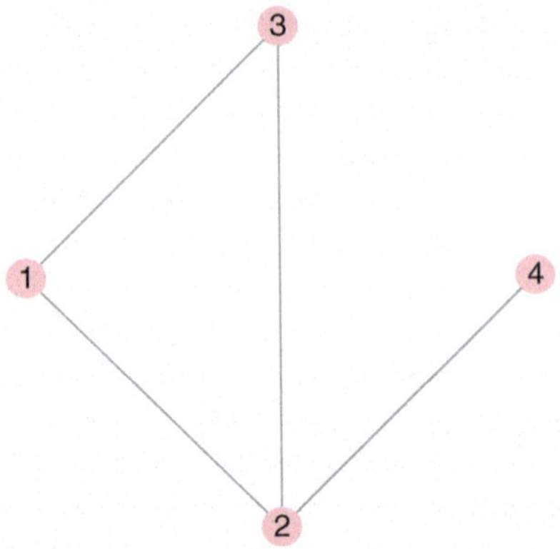

In the toy example given in Fig. 7, the number of edges is equal to 4 and the number of pairs of nodes is equal to $\frac{4\times 3}{2} = 6$ so the density is equal to $\frac{4}{6}$ 66.7 %.[7]

Because it is equal to the frequency of edges over the number of possible edges, the density is a measure of how densely connected the network is.

The "55-eqtl network" has 231 edges for 55 nodes; its density is thus equal to $\frac{231}{55\times 54/2}$ 15.6 %. It can be obtained with the function `edge_density`:

```
edge_density(glassoSubNet)
```

```
## [1] 0.1555556
```

It is expected that the density tends to decrease with the number of edges (see (Dorogovtsev and Mendes 2003) for examples of real-world networks together with their main characteristics).

Definition 2 (transitivity) The transitivity of a network is the number of triangles in the network divided by the number of triplets of nodes that are connected by at least two edges.

In the toy example given in Fig. 7, the transitivity is equal to $\frac{1}{3}$ 33.3 % (one triangle linking the nodes $\{1,2,3\}$ and three triplets with at least two edges: $\{1,2,3\}$, $\{2,3,4\}$ and $\{1,2,4\}$.

Speaking in terms of a social network, the transitivity thus measures the probability that two of my friends are also friends. A transitivity which is much larger than the density indicates that the nodes are not connected *at random* but on the contrary that there is a strong local connectivity (a kind of "modular structure"), which is often the case in real-world networks.

[7] The number of pairs for a set of n objects is equal to $\frac{n(n-1)}{2}$.

The "55-eqtl network" has a transitivity equal to 68.7 % that is obtained with the function `transitivity`:

```
transitivity(glassoSubNet)
```

```
##    [1]    0.6868448
```

As expected, the transitivity is much larger than the density for the "55-eqtl network" which shows a strong local connectivity.

4.3 Individual Characteristics

Once the network structure is analyzed globally, one may want to focus more precisely on nodes individually so as to extract the most "important" ones. Some simple numeric characteristics can be used to do so.

Definition 3 (degree) The *degree* of a node v_i is the number of edges adjacent to this node: $d_i = \left|\left\{(v_i, v_j) \in E : j \neq i\right\}\right|$.

Nodes that have a large degree are called *hubs*.

In the toy example given in Fig. 7, the degree of the node 2 is equal to 3 (three edges are afferent to node 2 linking node 2 to nodes 1, 3 and 4).

The degree is a measure of the node's "popularity." Using the function degree, the degrees of all nodes in the "55-eqtl network" can be obtained:

```
head(degree(glassoSubNet), n=5)
```

```
##    THRB    PSMC3IP    XIAP    ARHGAP8    X91721
##       5          1       3         18        16
```

The degree distribution of the "55-eqtl network" is shown in Fig. 8. This figure shows that most of the nodes have a very small degree (smaller than 5) whereas a few nodes have (comparatively) very large degrees (more than 20).

Many real-world networks are reported to have a *degree distribution* (i.e., the values $(\mathrm{P}(k))_k$ that counts the number of nodes with a given degree k) which fits a *power law*: $\mathbb{P}(k) \sim k^{-\gamma}$ for a given $\gamma > 0$. Thus, degree distributions are often displayed with log–log scales (i.e., $\log \mathrm{P}(k)$ versus $\log k$). In this case, a good linear fit indicates a power law distribution. The "55-eqtl network" is a bit too small to observe such a distribution but nevertheless, the degree distribution is skewed. Also, there is a higher proportion of nodes with a degree between 15 and 20. Looking at Fig. 9, we can see that this corresponds to the set of nodes that are highly connected.

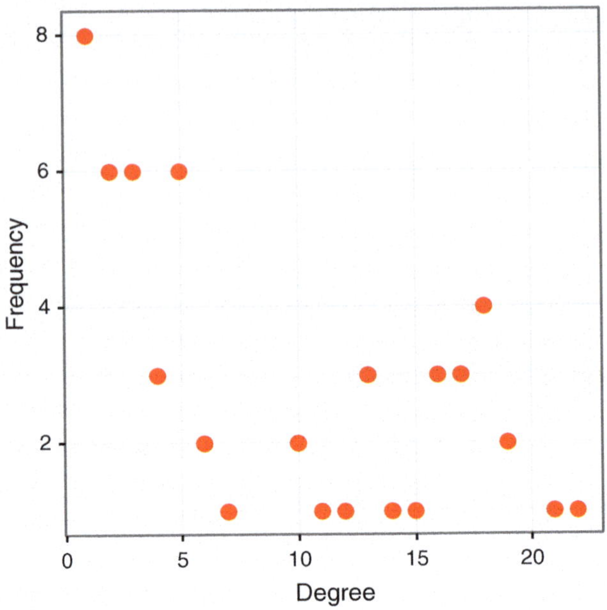

Fig. 8 Degree distribution for the "55-eqtl network"

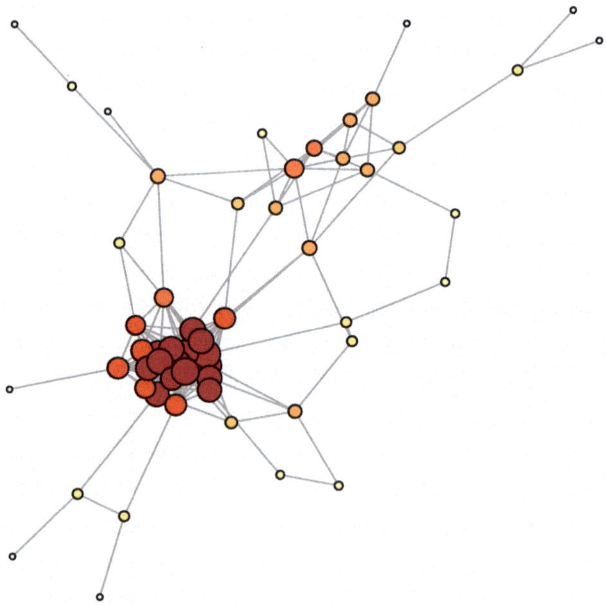

Fig. 9 "55-eqtl network": the node sizes and their colour intensities are proportional to their degrees

Definition 4 (betweenness) The *betweenness* of a node v is the number of shortest paths between any pair of nodes that pass through this node.

In the toy example given in Fig. 7, the betweenness of node 2 is equal to 2 because the shortest path between nodes 1 and 4 is $1 \rightarrow 2 \rightarrow 4$ and the shortest path between nodes 3 and 4 is $1 \rightarrow 2 \rightarrow 4$. All the other nodes have a betweenness equal to 0.

The betweenness is a centrality measure: nodes that have a large betweenness are those that are the most likely to disconnect the network if removed. They may thus correspond to genes of high importance. Using the function betweenness, the betweenness of the 55 nodes of the "55-eqtl network" can be obtained:

```
head(betweenness(glassoSubNet), n=4)
```

```
##      THRB   PSMC3IP      XIAP    ARHGAP8
##  137.41563  0.00000  57.47527   54.33676
```

The betweenness of every node is displayed in Fig. 10. It is interesting to note that nodes with high betweenness are not necessarily hubs. The nodes with the highest betweenness are more outside the set of nodes which are highly connected.

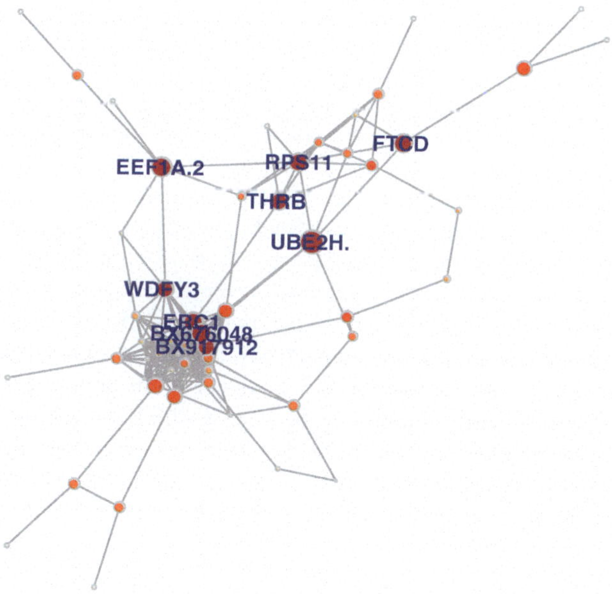

Fig. 10 "55-eqtl network": the node sizes and their colour intensities are proportional to their betweenness

4.4 Clustering

Clustering nodes in a network consists of partitioning the network into densely connected groups that we will call *modules* in the sequel. The nodes in a given module share a few number of edges (comparatively) with the nodes of other modules. Modules are often called *communities* in social sciences and *clusters* in statistics. A number of methods have been designed to address this issue and this section is much too small to go beyond scratching the surface of this topic. For further references on this topic, we advise the reader to refer to (Fortunato and Barthélémy 2007; Schaeffer 2007).

One of the most popular approaches for node clustering consists of maximizing a quality criterion called *modularity* (Newman and Girvan 2004):

Definition 5 (modularity) Given a partition $(\mathcal{C}_1,\ldots,\mathcal{C}_K)$ of the nodes of the graph, the *modularity* of the partition is equal to

$$\mathcal{Q}(\mathcal{C}_1,\ldots,\mathcal{C}_K)=\frac{1}{2m}\sum_{k=1}^{K}\sum_{v_i,\ v_j\in\mathcal{C}_k}\left(\mathbb{I}_{(v_i,v_j)\in E}-P_{ij}\right)$$

where $P_{ij}=\frac{d_i d_j}{2m}$, di the degree of node i and $m=|E|$ is the number of edges in the network.

In this definition, Pij plays the role of a probability to have an edge between vi and vj according to a "null model." In the "null model", the edges depend only on the degrees of each node and not on the clusters themselves: the larger the modularity, the more the edges are concentrated in the clusters $(\mathcal{C}_j)_j$. This model slightly differs from maximizing the number of edges in the clusters: edges that correspond to nodes with a large degree have a lesser impact in the modularity value: this aims at encompassing in the criterion the notion of *preferential attachment* (Barabási and Albert 1999), which is the fact that, in real networks, people tend to connect preferably with people who already have a large number of connections. Hence, the edges of very popular nodes (hubs) seem to be less "significant" (or, in other words, less important to define an homogeneous module). In particular, the modularity is known to better separate hubs (as compared to a naive approach consisting of minimizing the number of edges between clusters, that leads more frequently to have huge clusters and tiny ones with isolated nodes). Also, the modularity is not monotonous in the number of modules: it can thus be useful to decide on an adequate number of clusters. However, it is also known to fail to detect small modules (Fortunato and Barthélémy 2007). Several method can be used to find a partition that approximately optimizes the modularity.[8] In the R package igraph, several methods are implemented. In the following, we will use the function cluster_spinglass, which implements the method described in (Reichardt and Bornholdt 2006) (equivalent in certain cases to modularity optimization) and based on simulated annealing:

[8]The modularity maximization is an intractable problem which can be solved only for small networks. For large networks, fast algorithms are usually used to find an approximate solution.

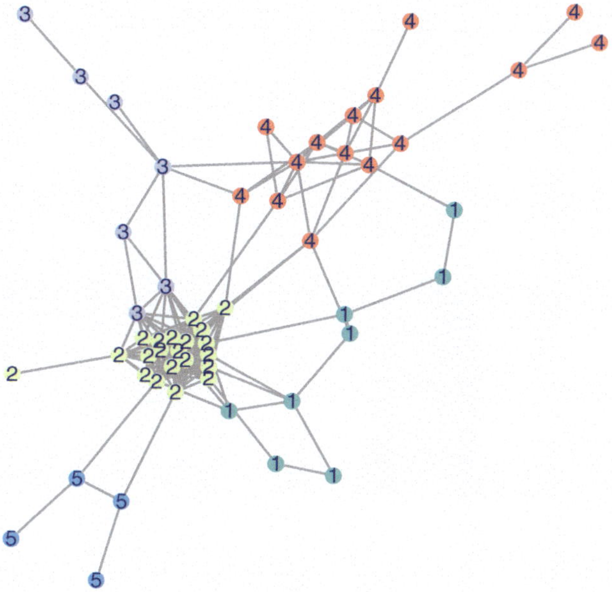

Fig. 11 Partition of the "55-eqtl network" into five modules. The colours and labels indicate module membership

```
finalClustering = cluster_spinglass(glassoSubNet)
modularity(finalClustering)
```

```
## [1] 0.3102359
```

```
head(membership(finalClustering))
```

```
##   THRB  PSMC3IP  XIAP  ARHGAP8  X91721  BX917912
##      4        4     5        2       2         2
```

```
sizes(finalClustering)
```

```
## Community sizes
##  1  2 3  4 5
##  8 21 7 15 4
```

Using this method algorithm,[9] the "55-eqtl network" could be partitioned into five modules (Fig. 11), of 8, 21, 7, 15 and 4 nodes, respectively. The modularity of this partition is equal to 0.31.

To assess if the modularity is significantly large (and hence if the partition is meaningful), a test of significance has been performed, as described in (Montastier et al. 2015; Rossi and Villa-Vialaneix 2011). This test is based on the computation of

[9] As the algorithm is partially stochastic, it has been run 100 times and only the best result has been kept.

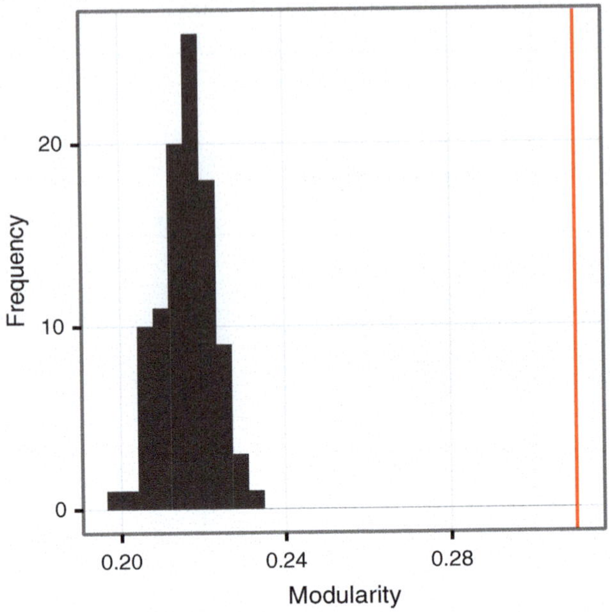

Fig. 12 Distribution of the maximum modularity over 100 random graphs with the same degree distribution as the "55-eqtl network" compared to the maximum modularity found for this network (*red vertical line*)

the maximum modularity for 100 random graphs with the same degree distributions as "55-eqtl network." The distribution of the maximum modularity for the random graphs is compared to the maximum modularity of the "55-eqtl network" in Fig. 12.

5 Biological Mining

Apart from providing easy-to-handle graphical displays, network analysis can be used forward to interpret the data. To that end, the analyst needs to go back to biological knowledge and extract coherent biological findings from statistical results. This analysis can be conducted in three steps:

1. Gene annotation
2. Biological enrichment
3. Biological networks

5.1 Gene Annotation

In the previous sections, expression data were used without taking into account the biological functions associated with nodes. Nodes are first a DNA sequence coming from RNA sequencing or probes on microarrays. According to the quality of the annotation of the studied genome, only part of the nodes are annotated. One of the advantages of a gene network is that all probes, even those that correspond to unannotated probes, can be used for the analysis whereas they are often left aside in other approaches. In the example of the greatest connected component of 55 nodes, 34 nodes were annotated in 2013 (Villa-Vialaneix et al. 2013) while 43 were annotated in 2015 thanks to the progress of the annotation of the pig genome.

Giving access to the original sequences is of prime importance when publishing transcriptomic data (see MIAME, Minimum Information About a Microarray Experiment (Brazma et al. 2001)). Data must be submitted to public repositories such as Gene Expression Omnibus (GEO, NCBI website)[10] or ArrayExpress (EMBL website)[11] and many others allowing the complete access to the probe sequence. At the time of publication, some related information may be associated with the sequence: current annotation with gene name or symbol, gene description, aliases, known orthologs, accession number of the sequence from which the probe has been designed.

Functional information could be associated to each gene product. A consortium tries to attribute functional terms with a curated approach (controlled vocabulary) named Gene Ontology (GO).[12] The biology is cleaved in three domains: Biological processes (e.g., glycolytic process), molecular function (e.g., acetyl-CoA transporter activity) and cellular component (e.g., glycosome). Other reliable functions may be obtained with KEGG (Kyoto Encyclopedia of Genes and Genomes).[13] KEGG is a database which gives access to many well-documented pathways such as signaling (e.g., PI3K-Akt signaling pathway), metabolism (e.g., lipid metabolism) or biological processes (e.g., cell growth and death).

Functional information for a full list of genes can be obtained from databases like DAVID (Database for Annotation, Visualization and Integrated Discovery)[14] with the downloadable application EASE (Expression Analysis Systematic Explorer)[15] or "Ensembl" with BioMart.[16] Care must be taken if an updated version is available. For instance, current annotation in Ensembl is the release 81—July 2015 at the time of this review. Also, the user has to carefully make the choice of the genome annotation to which to refer. For instance, for the pig genome, two genome annotations can be used: the one of the pig or the one of the human. At the date of this review, in BioMart:

- *Pig genome*: there are 18,466 Ensembl gene IDs (from 21,630) with at least one GO and a total of 180,197 GO term accessions. One gene is associated with 0 to 246 GO term accessions (the average is about 8 GO per Ensembl gene ID).

[10] http://www.ncbi.nlm.nih.gov/geo.

[11] https://www.ebi.ac.uk/arrayexpress.

[12] http://geneontology.org.

[13] http://www.genome.jp/kegg.

[14] https://david.ncifcrf.gov.

[15] https://david.ncifcrf.gov/ease/ease1.htm.

[16] http://www.ensembl.org/biomart/martview/79399dc2f5745752a66a5a4a43f32a38.

Table 1 Example of some systematic functional annotation for four genes out of the 43 that are annotated

Gene	GO Biological	GO Cellular	GO Molecular	KEGG
Symbol	Process	Component	Function	Pathway
ACBD5				
DECR2	Alcohol; metabolism	Peroxisome	Oxidoreductase activity	
ITGA8	Cell adhesion	Plasma membrane	Cell adhesion; molecule activity	
PDE8A	Cell	Insoluble fraction	Transition; metal	Purine
	Communication		Ion binding	Metabolism

These results were obtained with the EASE application

- *Human genome*: there are 20,632 Ensembl gene ID (from 22,699) with at least one GO and a total of 774,505 GO term accessions. One gene is associated to 0 to 1849 GO term accessions (the average is about 31 GO per Ensembl gene ID).

For genes in the same family, the gene annotation may be ambiguous between species, with possible false contributions to a function when using the human genome instead of the pig genome. However, using the human genome strongly increases the number of associated functions. For this reason, the human genome is preferred in the sequel, as a referenced mammalian genome. The lists of genes obtained from the different clusters obtained in Sect. 4.4 will be further studied. For instance, Table 1 shows an extract of some related functions for four of the 43 annotated genes. No functional information could be retrieved for the *ACBD5* gene while the *PDE8A* gene is much better annotated.

5.2 Biological Enrichment

Here, the reference genome for the pig species is the human genome in order to obtain richer biological information related to each gene. Another reliable step of the analysis of large transcriptomic data or of the clustering of co-expressed genes consists in identifying enriched biological functions associated with a set of selected genes.

Many free software (STRING,[17] GeneCodis,[18] WebGestalt[19] and DAVID[20] among others) or software under license, such as Ingenuity Pathway Analysis (IPA[21] and

[17] http://string-db.org.

[18] http://genecodis.cnb.csic.es.

[19] http://bioinfo.vanderbilt.edu/webgestalt.

[20] https://david.ncifcrf.gov.

[21] http://www.ingenuity.com/products/ipa.

others), are available to obtain enriched biological functions under different terms. The overall process is most often the same:

1. The first step is to attribute known biology terms for each gene from several databases (see Sect. 5.1). The most usual ones can be found below, but other reference databases may be more relevant to the studied species:
   - Gene Ontology.[22]
   - KEGG.[23]
   - Transcription factors[24] may give information about the transcription regulation of the targeted gene in the reference genome based on the known cis-regulatory element. This information could be particularly interesting with a co-expression analysis but must be used with care when dealing with data from homologous species.
   - Others, such as Omic Tools,[25] are useful for retrieving regulating miRNA or other non-coding RNA, common protein domain, co-cited in publications.
2. The second step is to identify the terms from the above lists and count the number of genes for each term (da Huang et al. 2009). A statistical test will then give the significance of the enrichment (Fisher's exact tests based on hypergeometric distribution (Fisher 1922) and correction for multiple testing (Benjamini and Hochberg 1995)).

With the 43 annotated nodes provided in this example, Webgestalt[26] recognized 40 unique genes with, e.g., "RNA transport" pathway significantly enriched (related to three nodes/genes, *PABPC1*, *EEF1A1*, *EEF1A2*). With GeneCodis,[27] co-occurrence findings are possible: three genes (*EEF1A1*, *NCOA2* and *THRB*) are significantly associated with "regulation of transcription, DNA-dependent (BP), nucleus (CC), protein binding (MF), V$MAZ_Q6" (transcription factor targets) meaning that the products of these three genes are localized in the nucleus with protein binding activity to regulate the transcription. The transcription factor MAZ (MYC-associated zinc finger protein (purine-binding transcription factor)) was demonstrated to be able to regulate the expression of these three genes.

In Table 2, from the 11 recognized genes (column "list size"), out of the 21 nodes of cluster 5 (see Fig. 11), two gene products (DECR2 and ACBD5, column "Support") are associated with a peroxisome localization in the cell. This function was said to be enriched compared to the 105 genes (column "Reference support"), which are localized in the peroxisome, among the 34,208 genes (column "Reference size") of the human genome. To evaluate this enrichment, a *p*-value based on

[22] http://geneontology.org.

[23] http://www.genome.jp/kegg.

[24] http://www.broadinstitute.org/gsea/msigdb.

[25] http://omictools.com/transcriptomics-c1178-p1.html.

[26] http://bioinfo.vanderbilt.edu/webgestalt.

[27] http://genecodis.cnb.csic.es/analysis.

Table 2 Enrichment analysis of the 21 nodes of the fifth cluster

Items	Details	Support	List size	Ref. support	Ref. size	*p*-value	Adj. *p*-value	Genes
GO:0006355	Regulation of transcription, DNA-dependent (BP)	3	11	1609	34,208	0.01290	0.01934	PDE8A, NCOA2, ERC1
GO:0005777	Peroxisome (CC)	2	11	105	34,208	0.00050	0.01462	DECR2, ACBD5
GO:0016020	Membrane (CC)	5	11	4065	34,208	0.00588	0.01763	TMEM126B, CCDC56, ERC1, ITGA8, ACBD5
GO:0016021, GO:0016020	Integral to membrane (CC), membrane (CC)	4	11	2933	34,208	0.01088	0.02177	TMEM126B, CCDC56, ITGA8, ACBD5
V$PAX4_03	V$PAX4_03	3	11	1033	34,208	0.00378	0.02267	ARHGAP8, ACBD5, MGEA5
GO:0016020	Membrane (CC)	5	11	4065	34,208	0.01588	0.04262	TMEM126B, CCDC56, ERC1, ITGA8, ACBD5
GO:0000139	Golgi membrane (CC)	2	11	420	34208	0.00769	0.04458	ERC1, B2M
GO:0007275	Multicellular organismal development (BP)	2	11	945	34208	0.03554	0.0485	ERC1, ITGA8

This result was obtained with GeneCodis and the Human genome as reference. To read the table, see the explanation in the text hypergeometric distribution (column "p-value") and its corresponding corrected p-value (column "ad$j$. p-value") were calculated.

5.3 Biological Networks

Biological networks can be constructed with free software like STRING (http://string-db.org) for functional association networks mainly based on Known and Predicted Protein–Protein Interactions but using also indirect (functional) associations (conserved co-expression data) or previous knowledge from literature.

Another software is Ingenuity Pathway Analysis (IPA), under license, not only allows the user to find enrichment for the called canonical pathways or biofunctions and others but also extracts biological networks based on all possible relationships across many databases and literature. IPA can propose networks with a limited total number of nodes (35, 70 or 140 nodes) including the best interactions between the input genes (in priority) and additional genes to obtain significant networks ranked with an associated score. Biological functions are associated with the proposed networks. In our example, cluster 2 contains 21 nodes, out of which five genes had an associated biological process enriched with GeneCodis and only 2 genes with Webgestalt. Only 50 % of the nodes were used to find associated biological functions because of the limitation of annotation and there was available biological information for about 10–35 % of the nodes.

The Ingenuity Pathway Analysis recognized the 11 annotated genes. IPA possessed a rich Ingenuity Knowledge Base with automated and manually curated information from all the databases presented before and also referenced all genes by possible gene interaction. Figure 13 shows the IPA network including all the 11 annotated genes of cluster 2. Associated functions are organismal survival (four genes), development (three genes), expression regulation (two genes). The colour code is related to the betweenness centrality of the node in the largest connected component before clustering (highest for *ERC1*). Figure 14 shows the network as displayed by Gephi[28] (Bastian et al. 2009) (this software easily imports graphs in graphml format as described in Sect. 4.4). The node size corresponds to the betweenness centrality and the colour intensity corresponds to the node degree, both restricted to the subgraph induced by the nodes in cluster 2.

Figures 13 and 14 correspond to two representations of the same cluster 2. The first one used the available biological information to propose an optimized network. The second one is built with the initial information on co-expression without prior biological knowledge. As observed in our previous work (Villa-Vialaneix et al. 2013), every cluster was associated with only one IPA network. In this case, 100 % of the annotated genes of cluster 2 are included in the same IPA network (it was only about 80 % for all clusters in our original work). Compared to the original paper (Villa-Vialaneix et al. 2013), it has to be noted that the initial annotation of *CCDC56* was changed into *COA3* (cytochrome c oxidase assembly protein 3) by IPA: both

[28] https://gephi.github.io.

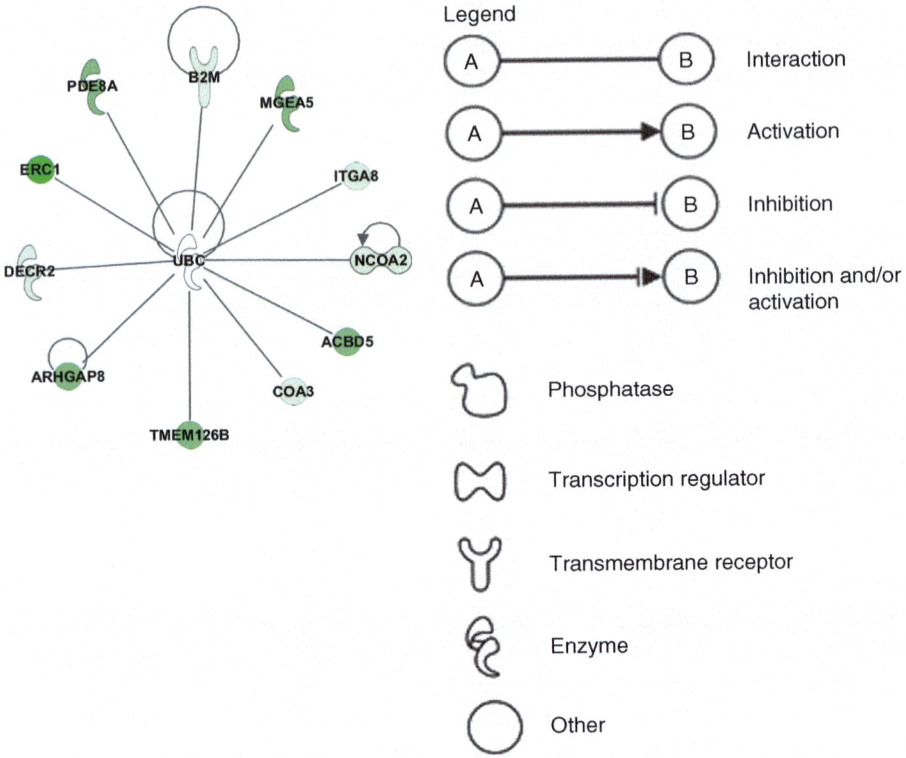

Fig. 13 IPA network including all the 11 annotated genes of cluster 2

names are indeed aliases. This simple example shows that a careful control of all the steps of functional annotation has to be performed. Finally, a biological hypothesis could be proposed for cluster 2, the density of which (0.74) is much higher than that of the entire network (0.15). Cluster 2 was found to correspond to the Ubiquitin Proteasome Pathway (see http://www.genome.jp/kegg-bin/show_pathway?hsa03050 for details) where the Ubiquitin protein binds most substrate proteins before their degradation by the proteasome.

These tools may be useful to help biologists to explore lists of genes or proteins coming from high-throughput technologies or lists coming from co-expression networks to explore associated functions with each community/cluster/module. However, the biologist must not forget his/her original biological question. In (Villa-Vialaneix et al. 2013), the aim was to identify key genes being regulated by a cis-eQTL and to underline possible important relationships between the original list of genes. Key genes could be unknown genes important from an eQTL point of view or important in the network. Such insights may encourage further biological analyses. Taken altogether, this complete set of tools

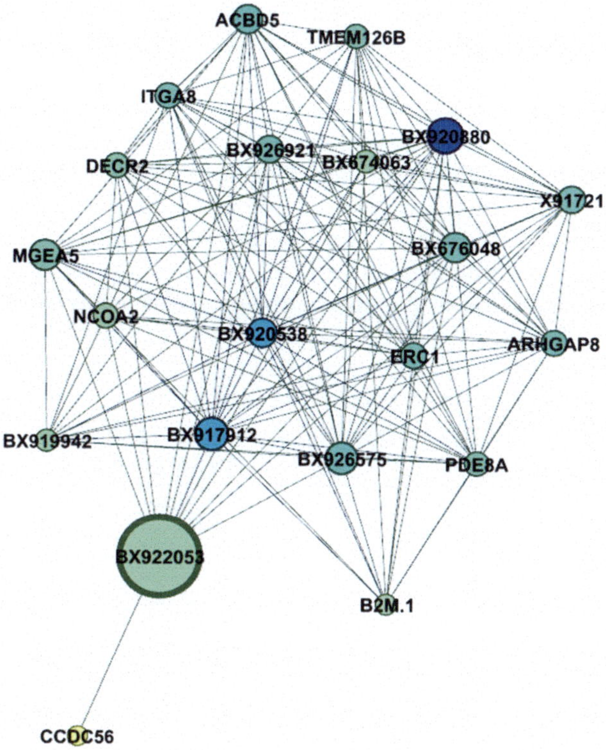

Fig. 14 Cluster 2 as displayed by Gephi

may be powerful to decipher the biological mechanisms and the genetics regulating the biology of a tissue and underlying complex traits of interest in an agronomic context.

6 Link with a Phenotype

Since an eQTL study is not a differential study, links of the genes with eQTLs and any phenotype are expected to be erratic a priori. In the pig example, let us consider the meat pH as a phenotype of interest: it is linked with meat quality. No high correlation was found between pH and gene expressions. A finer analysis is hence needed. The idea is to link the network structure with the phenotype of interest using spatial statistical tools. On average, are the genes of one cluster more correlated to the pH? Which genes are particularly correlated to the pH as well as their neighbouring genes on the network? Using spatial statistics, it is possible to detect modules and specific genes that are linked with a terminal phenotype. This analysis is not detailed in the present chapter and we encourage the interested readers to refer to (Villa-Vialaneix et al. 2013).

Conclusion

The prime objective was to decipher the processes underlying a list a genes whose expression is (partially) under genetic control. Due to an incomplete annotation of mammalian genomes, we proposed a statistical approach based on Gaussian graphical models for estimating and mining co-expression of a list of genes. This has led us to highlight a small subset of interesting genes (that are highly linked or central in the graph structure), and modules of densely connected genes. Roughly speaking, these modules were enriched in a single biological function, leading to a better clarity in the biological interpretation of the complex system under study. Last but not least, all these meaningful results are the consequence of joint work between statisticians and biologists, which proves the importance of the collaboration between the two fields.

References

Auber D (2003) Tulip: a huge graph visualisation framework. In: Mutzel P, Jünger M (eds) Graph Drawing Softwares, Mathematics and Visualization. Berlin, Heidelberg: Springer, pp 105–126

Barabási A, Albert R (1999) Emergence of scaling in random networks. Science 286:509–512

Bastian M, Heymann S, Jacomy M (2009) Gephi: an open source software for exploring and manipulating networks. In: E.e.a. Adar (ed) Proceedings of the Third International AAAI Conference on Weblogs and Social Media, pp 361–362. Menlo Park: AAAI Press. URL http://www.aaai.org/ocs/index.php/ICWSM/09/paper/view/154

Benjamini Y, Hochberg Y (1995) Controlling the false discovery rate: a practical and powerful approach to multiple testing. J Royal Stat Soc Series B 57:289–300

Brazma A, Hingamp P, Quackenbush J, Sherlock G, Spellman P, Stoeckert C, Aach J, Ansorge W, Ball C, Causton H, Gaasterland T, Glenisson P, Holstege F, Kim I, Markowitz V, Matese J, Parkinson H, Robinson A, Sarkans U, Schulze-Kremer S, Stewart J, Taylor R, Vilo J, Vingron M (2001) Minimum information about a microarray experiment (miame)-toward standards for microarray data. Nat Genet 29(4):365–371

Butte A, Kohane I (1999) Unsupervised knowledge discovery in medical databases using relevance networks. In: Proceedings of the AMIA Symposium, pp 711–715

Butte A, Kohane I (2000) Mutual information relevance networks: functional genomic clustering using pairwise entropy measurements. In: Proceedings of the Pacific Symposium on Biocomputing, pp 418–429

Csardi G, Nepusz T (2006) The igraph software package for complex network research. Inter J Complex Systems. URL http://igraph.sf.net

Dorogovtsev S, Mendes J (2003) Evolution of Networks. From biological Nets to the Internet and WWW. Oxford University Press

Dozmorov M, Giles C, Wren J (2011) Predicting gene ontology from a global meta-analysis of 1-color microarray experiments. BMC Bioinform 12(Supp 10):S14

Edwards D (1995) Introduction to graphical modelling. Springer, New York

Fisher R (1922) On the interpretation of x^2 from contingency tables, and the calculation of P. J Royal Stat Soc 85(1):87–94. doi:10.2307/2340521.JSTOR2340521

Fortunato S, Barthélémy M (2007) Resolution limit in community detection. In: Proceedings of the National Academy of Sciences, vol. 104, pp 36–41. doi:10.1073/pnas.0605965104; URL: http://www.pnas.org/content/104/1/36.abstract

Foygel R, Drton M (2010) Extended Bayesian information criteria for Gaussian graphical models. In: Proceedings of Neural Information Processing Systems (NIPS 2010), pp 604–612. Vancouver

Friedman J, Hastie T, Tibshirani R (2008) Sparse inverse covariance estimation with the graphical lasso. Biostatistics 9(3):432–441

Fruchterman T, Reingold B (1991) Graph drawing by force-directed placement. Software Pract Exp 21:1129–1164

Gillis J, Pavlidis P (2012) "guilt by association" is the exception rather than the rule in gene networks. PlLoS Computational Biology 8(3):e1002,444

da Huang W, Sherman B, Lempicki R (2009) Bioinformatics enrichment tools: paths toward the comprehensive functional analysis of large gene lists. Nucl Acids Res 37(1):1–13

Kogelman L, Zhernakova D, Westra H, Cirera S, Fredholm M, Franke L, Kadamideen H (2015) An integrative systems genetics approach reveals potential causal genes and pathways related to obesity. Genom Med 7:105. doi:10.1186/s13073-015-0229-0

Liaubet L, Lobjois V, Faraut T, Tircazes A, Benne F, Iannuccelli N, Pires J, Glénisson J, Robic A, Le Roy P, SanCristobal M, Cherel P (2011) Genetic variability or transcript abundance in pig peri-mortem skeletal muscle: eQTL localized genes involved in stress response, cell death, muscle disorders and metabolism. BMC Genom 12(548):548

Liu H, Roeber K, Wasserman L (2010) Stability approach to regularization selection (StARS) for high dimensional graphical models. In: Proceedings of Neural Information Processing Systems (NIPS 2010), vol. 23, pp 1432–1440. Vancouver. URL http://machinelearning.wustl.edu/mlpapers/papers/NIPS2010_0834

Lysen S (2009) Permuted inclusion criterion: a variable selection technique. Ph.D. thesis, University of Pennsylvania

Meinshausen N, Bühlmann P (2006) High dimensional graphs and variable selection with the lasso. Ann Stat 34(3):1436–1462

Montastier E, Villa-Vialaneix N, Caspar-Bauguil S, Hlavaty P, Tvrzicka E, Gonzalez I, Saris W, Langin D, Kunesova M, Viguerie N (2015) System model network for adipose tissue signatures related to weight changes in response to calorie restriction and subsequent weight maintenance. PLoS Comput Biol 11(1):e1004,047. doi:10.1371/journal.pcbi.1004047. First co-author

Newman M, Girvan M (2004) Finding and evaluating community structure in networks. Phys Rev E 69:026,113. doi:10.1103/PhysRevE.69.026113.URL, http://www.citebase.org/abstract?id=oai%3AarXiv.org%3Acond-mat%2F0308217

Reichardt J, Bornholdt S (2006) Statistical mechanics of community detection. Phys Rev E 74(016110)

Rossi F, Villa-Vialaneix N (2011) Représentation d'un grand réseau à partir d'une classification hiérarchique de ses sommets. Journal de la Société Française de Statistique 152(3):34–65. URL http://publications-sfds.math.cnrs.fr/index.php/J-SFdS/article/view/82/73

Schaeffer S (2007) Graph clustering. Comp Sci Rev 1(1):27–64

Schäfer J, Strimmer K (2005a) An empirical bayes approach to inferring large-scale gene association networks. Bioinformatics 21(6):754–764. doi:10.1093/bioinformatics/bti062

Schäfer J, Strimmer K (2005b) A shrinkage approach to large-scale covariance matrix estimation and implication for functional genomics. Stat Appl Genet Mol Biol 4:1–32

Shannon P, Markiel A, Ozier O, Baliga N, Wang J, Ramage D, Amin N, Schwikowski B, Ideker T (2003) Cytoscape: a software environment for integrated models of biomolecular interaction networks. Genome Res 13(11):2498–2504

Villa-Vialaneix N, Liaubet L, Laurent T, Cherel P, Gamot A, San Cristobal M (2013) The structure of a gene co-expression network reveals biological functions underlying eQTLs. PLoS One 8(4), e60,045. doi:10.1371/journal.pone.0060045

Applications of Systems Biology to Improve Pig Health

Martine Schroyen, Haibo Liu, and Christopher K. Tuggle

Abstract

In the pig, there are thus far only a handful of examples of health/disease studies approaching a systems biology level analysis, and this is in sharp contrast to the substantial amount of published porcine data on whole genome, transcriptome and proteome experiments with regard to economically important swine diseases. However, systems biology is very powerful since it attempts to understand how these distinct -ome parts work together to create emergent properties that are less likely to be recognized in the analysis of only one component of the system. By integration of the different -omics datasets, systems biology tries to create a more complete understanding of the observed immune response. Until now, such integrative analyses are still in their infancy in terms of application to pig health.

In this chapter, we will cover systems biology tools for network analyses and multilevel data integration, and give examples of their implementation in pig disease studies. Next, we will discuss the need for visualization to interpret the vast amount of data created in -omics studies. Furthermore, the upcoming use of bloodomics is described, since blood is a very relevant immune-related tissue and biomarkers in the blood can easily be assessed and implemented in selection strategies. We conclude with specific examples of -omics and initial systems biology methods on viral (PRRSv) and bacterial (*Salmonella*) infections, since both agents are economically important pathogens causing disease in pigs and substantial genomics analyses on the response to these pathogens have been conducted to date. In the future, forthcoming consortia such as the FAANG project will accelerate our ability to apply systems biology tools to improving pig health.

M. Schroyen • H. Liu • C.K. Tuggle (✉)
Department of Animal Science, Iowa State University,
2255 Kildee Hall, Ames, IA 50011, USA
e-mail: cktuggle@iastate.edu

H.N. Kadarmideen (ed.), *Systems Biology in Animal Production and Health, Vol. 2*,
DOI 10.1007/978-3-319-43332-5_2

1 The Time Is Right to Apply Genomic Tools for Improvement of Complex Health Traits in Pigs

In the last few decades, growth, meat quality, as well as feed and reproduction efficiency have been the most well-studied traits in swine breeding; however, in recent years, pig performance in the face of disease challenge is becoming progressively more important. Hence, selection objectives in the swine breeding industry have broadened to include traits that reflect overall robustness and disease resistance (Mellencamp et al. 2008). Heritability of cellular immune traits associated with resistance are often very high (Flori et al. 2011), so genetic selection towards more resistant pigs is certainly a feasible method to improve both animal production and welfare, but the possible existing trade-off with other traits should be kept in mind (Rauw 2012; Stear et al. 2001). However, the biology behind resistance towards even a single pathogen is highly complex and dynamic, creating an opportunity to apply systems immunology or systems biology to improve disease resistance (Kidd et al. 2014). The integration of experimental and computational research would allow a better understanding of these complex biological systems (Hollung et al. 2014), and high-throughput technologies, measuring thousands of parameters at once, would provide the requisite datasets (Kidd et al. 2014). To date, substantial whole genome, transcriptome and proteome data have been collected with regard to several economically important swine diseases; metabolome and microbiome datasets are also growing. The biggest challenge lies in bringing the data together to understand the immune responses in a comprehensive way and to use such information to improve pig health practically and sustainably.

2 Systems Biology Tools and Their Use in Pig Disease Studies

A multitude of pig disease genetics studies make use of knowledge gathered by genome-wide association studies (GWAS) through examining possible associations between single nucleotide polymorphisms (SNPs), insertions, deletions or copy number variants (CNVs) and the disease of interest (Arakawa et al. 2015; Fowler et al. 2013; McKnite et al. 2014; Sharma et al. 2015). When a mutation is found to be associated with the disease trait, one can select for it, with consideration of additional effects the mutation might have on other traits. At the transcriptomic level, differential expression (DE), usually over time or between disease states, can be informative. When the expression level of genes that differ between diseased and healthy phenotypes can successfully be repeated in other populations, they can be used as biomarkers. This is called signature-based analysis (Bebek et al. 2012). However, a single marker or a set of marker genes is usually not enough to explain or predict a complex phenotype. Integrative analyses merging gene expression profiles with pathway data have been shown to be helpful in understanding immune responses (Sahadevan et al. 2014). The gene set enrichment analysis (GSEA) algorithm is a powerful method to find enriched pathways in the transcriptome

(Subramanian et al. 2005). With this method, a pathway is scored according to how many and how enriched the genes representing the pathway are in the extreme up-regulated or down-regulated lists of genes. A similarly well-known annotation tool is the Database for Annotation, Visualization and Integrated Discovery (DAVID) (Huang da et al. 2009), which works together with the Gene Ontology (GO) and the Kyoto Encyclopedia of Genes and Genomes (KEGG) pathway databases (Kanehisa et al. 2014) for pathway visualization. InnateDB can be a powerful tool to annotate sets of genes to specifically examine the innate immune response (Lynn et al. 2008). One or more of these tools are typically used when analyzing pig transcriptomic data from microarray or RNA-seq. A few drawbacks in these pathway-based analyses are that well-characterized pathways are easier to find than those that describe the function of less studied genes and may be overemphasized in such annotation analyses. There is also the assumption that expression patterns of genes coding for proteins in a pathway should show a clear correlation with others in the same pathway, which is not necessarily always the case (Bebek et al. 2012).

However, systems biology goes further than a genome or transcriptome study and the annotation of overrepresented and underrepresented pathways. It is a combination of knowledge concerning several biological system parts, e.g. DNA, RNA, proteins, cells, tissues, organs, organisms and ecologies. Rather than producing these data and solely giving a summation of the outcome in each field, systems biology attempts to understand how the parts work together to create emergent properties that are less likely to be observed (recognized) in analyses of only components of the system. A further goal, specifically for systems immunology, is predicting how, in the light of health research, genetic and regulatory interactions, as well as environmental factors, orchestrate responses to disease (Tuggle et al. 2011). In what follows, we describe network-based and multilevel data integration analyses and, although both methods are still in their infancy in pig disease studies, we provide examples and illustrate their use.

2.1 Network-Based Analyses of Porcine Immunological Responses

One emerging network-based tool is the Weighted Gene Co-expression Network Analysis (WGCNA) developed by Langfelder and Horvath (2008). WGCNA was originally developed for microarray analyses, but is also applicable on RNA-seq data (Langfelder and Horvath 2008). Whereas in DE analyses only genes that pass an arbitrarily defined statistical threshold for DE are used for analysis, in WGCNA genes with a similar expression pattern across the experiment are clustered together into modules. Thus, genes with only a small but consistent difference over time or between phenotypic groups can be clustered in a module and will be considered. A unique and useful component of the WGCNA package is that, after clustering, the calculated eigengene of a module (defined as the first principal component of that module) can be correlated with an external numerical or categorical trait. Correlation coefficients together with nominal *p*-values indicate the strength of a module's relationship to the trait of interest. To understand their biological relevance to the trait, the genes in a module can then together be analyzed with GO annotation tools.

With regard to disease in pigs, Kommadath et al. (2014) used WGCNA to examine the blood transcriptome in pigs infected with *Salmonella enterica* serovar Typhimurium (ST) and grouped in extremes for the amount of fecal shedding bacteria, e.g. eight low shedders versus eight high-shedding animals (for more details, see Section 4.6). Four modules were correlated with shedding, two of which were annotated for immune functions and many of the immune genes in these modules were up-regulated 2 days post-inoculation. Some of the genes found by this method were already known to be related to a *Salmonella* infection such as *SLC11A1*, *TLR4*, *CD14* and *CCR1*. For others, such as *SIGLEC5*, *IGSF6* and *TNFSF13B*, the association with *Salmonella* shedding was novel (Kommadath et al. 2014). In a PRRS microarray study comparing four phenotypic groups of animals with extremely different growth rates and viremia levels after a PRRS virus (PRRSv) infection, limited information was obtained through linear modeling of blood gene DE that contrasted pigs with these extreme phenotypes. However, when using WGCNA, an interesting immune-related module was found containing cytokines, chemokines, interferon type I stimulated genes, apoptotic genes and genes regulating complement activation. The eigenvalue of this cluster for each pig's data correlated both with weight gain (WG) after 42 days post-infection with PRRSv and the WUR10000125 (WUR) SNP genotype on *Sus scrofa* chromosome 4 (SSC4), which explained a large proportion of the genetic variance for viral load and, to a lesser extent, weight gain (Boddicker et al. 2012). The genes in this WGCNA module could be useful targets for further selection against PRRS resistance (Schroyen et al. 2015). For more details, see Section 4.2 entitled "Transcriptomic Analysis of Host Response to PRRSv".

Although not directly relevant to pig health, but using the pig as a model for human health, Kogelman et al. (2014) applied WGCNA on RNA-seq of subcutaneous adipose tissue from 36 pigs with different risk levels for obesity. The module that showed the highest correlation with obesity-related traits contained 275 genes. The most significant GO term defining this cluster was "osteoclast differentiation" and osteoclasts are derived from macrophages, an immune cell type highly up-regulated in obese individuals. Other immune-related GO terms enriched in this gene list involved natural killer cells and B cell receptor signaling pathways, enlightening the association between obesity and immune-related complications (Kogelman et al. 2014).

Partial Correlation and Information Technology (PCIT) (Koesterke et al. 2013, 2014; Reverter and Chan 2008), together with the regulatory impact factor (RIF) and phenotypic impact factor (PIF) algorithms (Reverter et al. 2010) were also used to examine differences in networks drawn from different biological states. With PCIT, the co-expression correlation between each gene pair in a network is calculated and changes between different phenotypic groups are noted. RIF and PIF algorithms compute differential wiring between nodes for different treatments or groups to identify novel regulators. Using PCIT, Schroyen et al. (2015) found tighter connections to genes in the immune activation pathways in the low weight gain group compared to the high weight gain group after PRRS infection, indicating that one of the most significant differences between these two phenotypic groups was an immune network response. However, when comparing WGCNA and PCIT results, the WGCNA method seems to be more sensitive, since the PCIT algorithm removes edges (gene interaction measures), which can sometimes lead to an underestimation of the importance of a hub gene, and has consequences for biological interpretations (Kadarmideen and Watson-Haigh 2012).

2.2 Multilevel Data Integration Analyses of Pig Disease Biology Are Sparse

To date, there are only a few examples of integration of multiple -omics datasets in research on pig disease. Most common data integration strategies are comparisons between the transcriptome and miRNAome, since the correlation analysis of the mRNA transcriptome and miRNAome data can reveal and explain the control of reciprocal expression patterns of predicted target mRNAs. An example is the negative correlation between expression levels of miRNAs and their predicted target genes in the swine leukocyte antigen (SLA) complex region found when comparing mRNA-seq and miRNA-seq data for liver, longissimus dorsi and abdominal fat (Endale Ahanda et al. 2012). The SLA region was chosen since this region is highly associated with immune response traits in pigs, for instance, in case of infectious diseases or after vaccination (Lunney et al. 2009), and miRNAs can play a crucial role in fine-tuning this immune response. With TargetScan, PACMIT and TargetSpy, several polymorphic miRNA target sites were found and SNPs in these 3′ untranslated regions (3′-UTR) were predicted to lead to altered miRNA regulation patterns (Endale Ahanda et al. 2012).

Bao et al. (2014) examined the buffering capacity of miRNAs in response to a *Salmonella* infection, i.e. the ability to lower the expression variation of target mRNAs, rather than changing their expression level. A significant buffering capacity was seen in lowly to moderately expressed target mRNAs when compared to non-target mRNAs, but this difference was not seen for highly expressed genes. In response to infection, at 2 days post-infection (dpi) in both up-regulated and down-regulated genes, an additional buffering capacity was noticed for the target mRNAs, which was not the case for the non-target mRNAs. This result was interpreted as indicating that such miRNAs cause the existing transcriptional network to rewire more tightly after infection (Bao et al. 2014). Other examples of miRNA–mRNA comparisons in pig disease studies will follow in the example sections of this chapter.

Another example of multilevel data integration analysis can be seen in the combination of GWAS and transcriptomic data. The first expression quantitative trait loci (eQTL) studies in pig were conducted to examine muscle development, carcass and meat quality traits; however, more and more eQTL studies have focused on pig health (Ernst and Steibel 2013). Ponsuksili et al. (2012) investigated the relation between SNP markers from the PorcineSNP60 BeadChip, gene expression in liver and muscle measured with an Affymetrix porcine genome array and plasma cortisol levels, which is important in regulating immune function. They used the network edge orienting (NEO) R software package to predict causal interaction between the three datasets and found 26 and 70 candidate genes in liver and 2 and 25 candidates in muscle to affect or respond to plasma cortisol levels, respectively (Ponsuksili et al. 2012). Chomwisarutkun et al. (2013) used a custom-designed microarray targeting previously detected QTL regions to find candidate genes for inverted teat defects as opposed to an earlier study which used a commercially available array. They found a number of DE genes in both epithelium and mesenchyme, almost all belonging to cell signaling pathways and encoding many members of the signaling cascades of growth factors (Chomwisarutkun et al. 2013). Reiner et al. (2014) used an Affymetrix porcine genome array and found 193 *cis*- and *trans*-eQTL, including 55 eQTL in a functional hotspot on SSC13, and they identified several candidate genes for a genetic predisposition for susceptibility to *Actinobacillus pleuropneumoniae*. With the increase of RNA-seq data, it has now become quite easy to assess allele-specific expression in heterozygous individuals. For an example on PRRS and allele-specific expression, we refer to the study done by Koltes et al. (2015) described below (see Section 4.1).

3 Visualization Tools Improve Our Ability to Identify and Interpret Complex Relationships

With the increase of -omics data and the complexity of data analyses, data visualization is becoming fundamental for the interpretation of high-dimensional molecular interactions. Tools to visualize GO enrichment analysis results, such as Gorilla (Eden et al. 2009), AmiGO (Carbon et al. 2009), Panther (Mi et al. 2013), REVIGO (Supek et al. 2011) and others, are freely available. In addition, there are also a few network expression tools available. One well-known tool to visualize large datasets is Cytoscape (Shannon et al. 2003). Cytoscape is an open source software platform that easily can be customized with plug-ins and shows data as nodes and edges in a network to which multiple levels of annotation can be added and in which genes can be selected or filtered out. Another freely available program is BioLayout Express^{3D} (BE3D), which draws co-expression networks (Freeman et al. 2007). A Pearson's correlation coefficient threshold decides which genes (nodes) are kept for visualization and a Markov clustering algorithm defines genes with similar expression patterns into clusters. Within BE3D are numerous user-defined variables for displaying these clusters, including the ability to label nodes with any user-inputted variable. For example, it is possible to overlay onto a gene expression-based network a visualization of correlation of such expression to an external trait such as pathogen level or growth during infection for the pigs in the study.

Kapetanovic et al. (2013) analyzed the expression profiles of pig alveolar macrophages (AM), bone marrow-derived macrophages (BMDM) and monocyte-derived macrophages (MDM) at 0 and 7 h after LPS stimulation. After stimulation, the expression profiles of AM were clearly distinct from those of BMDM and MDM, indicating a different regulation of LPS-stimulated genes in these macrophages. They also used the tool to compare expression patterns after stimulation between human, mouse and pig macrophages and showed clusters of genes with up-regulated expression patterns in human and pig that were not up-regulated in mouse macrophages or vice versa (Kapetanovic et al. 2013). It is even possible to use tissue-specific expression patterns from microarray data from many tissues obtained from healthy pigs to visualize the relationships of immune cells and their expression patterns versus other cell types (Freeman et al. 2012).

In Schroyen et al. (2016), BE3D identified clusters of genes whose expression patterns measured by RNA-seq differed between susceptible and more resistant animals in response to PRRS according to the WUR SNP, which will be described below (see Section 4.1). One cluster of 516 transcripts showed an apparent dissimilarity between the two contrasted groups and could be linked to signaling pathway differences involved in viral entry and replication.

Another example of the successful use of BE3D was described by the immune response annotation group (IRAG) (Dawson et al. 2013). IRAG was able to improve the characterization of the pig immunome by using correlation network analyses of transcriptomic data. In this massive study, genes were clustered according to their expression patterns in blood macrophages and lymph nodes derived from a multitude of pig stimulation, infection and disease studies. A cluster of 619 probesets, representing at least 511 transcripts, was significantly enriched for human immune-related GO terms. Since only 16% of these genes had been annotated in the pig, evidence was provided for the involvement of over 500 genes in immune responses that had not previously annotated for function in immune response processes (Dawson et al. 2013).

4 Bloodomics

More and more studies aiming to genetically improve livestock's robustness involve whole blood to define the immune capacity or immunocompetence of an individual to different stimuli (Mach et al. 2013) and potentially identify predictive biomarkers for resistance or resilient pigs (Huang et al. 2011). The term "bloodomics" encompasses all molecular profiling -omics tools that have been applied to peripheral blood, in which the blood transcriptome plays an influential role (Mohr and Liew 2007). For the immune system, blood is a very relevant tissue, since cells of the immune system circulate between central and peripheral lymphoid organs as well as migrate to and from sites of injury via the blood (Chaussabel et al. 2010). Whereas in 2002, very few blood transcriptomic studies were executed on any animal species, by 2014, a significant number of studies based on the blood transcriptome had been published on several animal species, and in particular for cattle and pigs as livestock species (Chaussabel 2015; Schroyen and Tuggle 2015).

Whole blood studies have several advantages such as the ease of collection and the repeated sampling of the same individual during response to a stimulus, which allows accurate within-animal comparison back to the baseline prior to infection.

Examining whole blood also facilitates the ability to develop a genetic marker screening based test, which should be relatively easy to obtain on a large scale in a commercial setting given that blood sampling is a common surveillance method in veterinary practice. Genes expressed in peripheral blood cells have been shown to reflect molecular mechanisms underlying differences in production traits and it can be an easily accessible source of information when monitoring physiological changes (Jegou et al. 2016). The genetic blood markers could include total and differential white blood cell counts, peripheral blood mononuclear leukocyte subsets and acute phase proteins, specific and non-specific antibodies, cytokines, as well as a set of differentially expressed genes between a healthy and diseased status. In Clapperton et al. (2009) and Flori et al. (2011), sets of porcine immune trait markers that can be used for selection, together with their heritability coefficients, are listed. However, since whole blood comprises a varying number of cell types, gene expression and protein differences from sample to sample should be interpreted with great caution. Gene expression patterns are highly dependent on the composition of the underlying cell population. Knowledge on immune cell specific expression could help with the investigation of exactly which cells are activated (Abbas et al. 2005). Computational methods such as cell type enrichment analysis (CTEN) (Shoemaker et al. 2012) or the tissue expression module in the annotation tool DAVID, used effectively by Hulst et al. (2013), could give an idea of the cell types dominating the whole blood transcriptome/proteome response. Complete blood counts (CBCs) as a covariate in statistical analyses can be adjusted for such differences across replicate blood samples. Furthermore, with such CBC data, the transcriptional response data can be deconvoluted to help identify the unique regulatory control of specific cellular responses to pathogens (Shen-Orr et al. 2010).

As with systems biology in general, one of the current hurdles with the interpretation of data from blood transcriptomic research is the organization of the data and the integration of different components such as sample information, quality of data, clinical information collected at the time of sampling and results of other cellular and molecular platforms (Chaussabel et al. 2010).

Example 1: Overview of -Omics Studies Concerning Porcine Reproductive and Respiratory Syndrome (PRRS) in Pig

Porcine reproductive and respiratory syndrome (PRRS), also known as mystery swine disease or blue ear disease, emerged in the late 1980s and 1990s and is to date one of the most economically important diseases affecting pigs worldwide (Holtkamp et al. 2013; Zimmerman 2003). The disease is caused by a single-stranded RNA virus belonging to the Arteriviridae family and, as its name reflects, affects two branches of the pig breeding industry. On the one hand, there are severe reproduction losses due to infertility, late-term abortions and mummified and stillborn fetuses. On the other hand, grower-to-finisher pigs suffer from serious pneumonia, which leads to increased pig morbidity and mortality rates (Rossow 1998). Depressed growth rates in subclinical infections are also significant, and to date production costs are estimated at $664 million a year, and that is only for the USA (Holtkamp et al. 2013). It is therefore not surprising that many efforts were made to

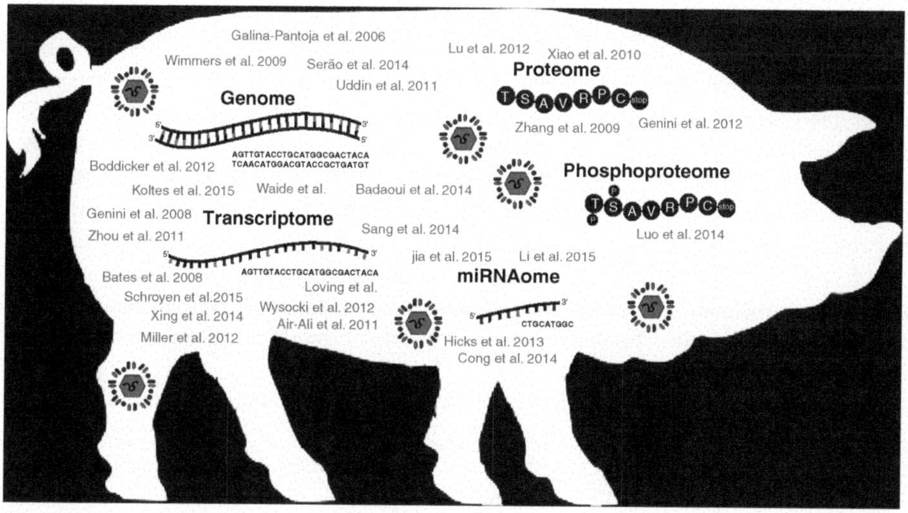

Fig. 1 Overview of -omics studies concerning porcine reproductive and respiratory syndrome (PRRS) in pig. For more details, see "Example 1: Overview of -omics studies concerning porcine reproductive and respiratory syndrome (PRRS) in pig"

understand PRRSv and its replicative life cycle, but the host point of view during PRRSv infection is also extensively studied. In this section, we give an overview of the different host-related -omics studies performed (Fig. 1) and, whenever present, the systems biology approaches utilized.

4.1 Linking Host Genomic Variation to Responses to PRRS

The first studies on host genetic variation associated with variation in response to PRRS used a limited set of SNPs. Galina-Pantoja et al. (2006) examined the association of phenotypes with 60 SNPs targeting host genes known to be associated with virus replication and viral entry into cells, as well as genes for receptors, macrophage and other innate immunity functions. They showed that in sows before and after infection with the virus, several of the SNPs tested were found to be associated with reproductive traits such as number of piglets born alive (Galina-Pantoja et al. 2006); these experiments were also summarized by Mellencamp et al. (2008). However, resistance is a complex and polygenic trait with substantial environmental influences; therefore, it is clear that selecting the best DNA marker or the best marker combination is complicated. Markers have to be consistent across datasets and they must have a positive effect on multiple traits and not be favorable for some and detrimental for others. Wimmers et al. (2009) used 88 markers, including 72 microsatellites and 16 biallelic markers, to find loci controlling the immune responsiveness in grower-to-finisher pigs. They screened for quantitative trait loci (QTL) by measuring complement activity, acute phase response and antibody response in animals before and after vaccination against *Mycoplasma hyopneumoniae*, herpesvirus I and PRRSv. In total, 21 QTLs were detected with a genome-wide significance level of 1%. These QTLs harbor several candidate genes for the traits examined (Wimmers et al. 2009). Uddin et al. (2011) used a panel of 79 microsatellites and 3 biallelic markers to search for immune-related QTLs. As innate immune traits they measured interleukin 2 (IL2), IL10, interferon gamma (IFNG), Toll-like receptor 2 (TLR2) and TLR9 levels in serum before and after vaccination with *M. hyopneumoniae*, PRRSv or tetanus toxoid (Uddin et al. 2011). The five traits were influenced by earlier described and newly found QTL on multiple chromosomes, implying multiple genes involved. Several candidate genes contributing to immune function were proposed for the three different vaccination experiments (Uddin et al. 2011).

However, although such analyses do help to discover regions containing QTL of interest, denser marker sets such as the porcine 60 K SNP chip could fine map the underlying genetic basis for these immune responses. However, substantially larger datasets are needed for such analyses. Serão et al. (2014) used the porcine 60 K SNP chip to perform a GWAS in a sow herd ($n = 641$) before and after a PRRS outbreak. They found a number of genomic regions strongly correlated with number of stillborn piglets, number and percentage of born dead piglets and sample-to-positive antibody ratios during and/or before PRRS infection. SNPs in these regions were found near genes associated with reproductive performance or immune response (Serão et al. 2014). Boddicker et al. (2012) also used this 60 K SNP chip, but focused on grower-to-finisher pigs and their genomics in relation to PRRSv infection. They found the QTL on SSC4 harboring the WUR SNP marker that has been associated with WG as well as PRRSv viremia levels, as described above (Boddicker et al. 2012). The effect of the SSC4 region and of WUR in particular was successfully validated in additional trials on animals with a different genetic background (Boddicker et al. 2013, 2014). This WUR marker maps close to several members of the guanylate binding protein (GBP) family which are known to be induced by gamma interferon. A transcriptomic approach was performed to identify differential expression between pigs with alternate QTL genotypes and potentially elucidate the underlying causal mutation. Koltes et al. (2015) specifically examined the expression of all genes in the region with high linkage disequilibrium to the WUR marker and determined that *GBP5* was differentially expressed between WUR genotype groups. Through deeper analysis of the RNA-seq data, they found a putative causal mutation causing differential splice variants of *GBP5*.

However, although these genomic analyses could lead to SNPs with large effects on phenotypes or even discover causal mutations, and the pig breeding industry could use them for selection towards better performing animals, such analyses often give little or no information about the molecular mechanisms that underlie these differences in phenotypes. In an integration of SNP association data with genome functional annotation, Waide et al. (submitted) performed GO enrichment analyses on sets of genes in close vicinity of SNPs associated with viral load and weight gain. They analyzed gene sets located within 250 kb of

SNPs that were associated with these traits ($-\log_{10}(p\text{-value}) > 2.5$). Analyses were performed using Panther (Mi et al. 2013) on a total of 13 trials of approximately 200 animals per trial and infected with the KS06 or NVSL PRRSv strain (Waide et al. submitted). For the SNPs associated with viral load, enriched biological processes (BP) terms for the KS06 strain included natural killer cell activation, immune response and B cell-mediated immunity, although the latter was not significantly enriched after Bonferroni correction. For the NVSL strain, enriched BP terms were immune response, metabolic process and lysosomal transport. For the SNPs associated with weight gain, antigen processing and presentation via MHC class II was the most enriched BP GO term for KS06; however, after Bonferroni correction, this term was no longer significant. Hence, it is possible to find groups of genes predicted to have functional differences between pigs with extreme phenotypes while using genomic rather than transcriptomic data. Since there are a large number of GWA studies available, it might be worthwhile to apply this approach to other existing datasets.

4.2 Transcriptomic Analysis of Host Response to PRRSv

Without doubt, the majority of research on host response to PRRSv is performed on the transcriptomic level. At the beginning of the twenty-first century, a multitude of microarray studies were performed examining host response to PRRSv, and these mostly in porcine alveolar macrophages (PAMs) (Genini et al. 2008; Zhou et al. 2011), lung (Bates et al. 2008; Xing et al. 2014), bronchial lymph nodes (Bates et al. 2008) and blood (Schroyen et al. 2015; Wysocki et al. 2012). Some of these studies compared non-infected with infected cells or tissues, while others focused on breed-specific (Ait-Ali et al. 2011; Xing et al. 2014) or within-breed resistance differences after infection (Boddicker et al. 2014). At the present time, the first RNA seq studies on host response to PRRS have been reported (Badaoui et al. 2014; Koltes et al. 2015; Miller et al. 2012; Sang et al. 2014; Schroyen et al. 2016). These RNA-seq studies examined blood, macrophages and tracheobronchial lymph nodes. Differentially expressed genes were often annotated as pro-inflammatory and several signaling pathways linked to the innate immune response surfaced. Overall, it has been shown that the PRRS virus triggers an atypical innate immune response, with less type I interferon α (IFNα) production compared to other viruses (Van Reeth et al. 1999), which leads to a reduced expression of interferon-induced genes and pathways. Better performing animals, that are less affected by viral infection, are believed to trigger their immune system earlier and possibly have a more effective response than the more susceptible animals, as seen by the expression profile differences (Ait-Ali et al. 2011; Schroyen et al. 2015), as well as when comparing cytokine levels in the sera (Souza et al. 2013; Van Reeth et al. 1999). The earlier described BE3D analysis of all available Affymetrix data on porcine immune response (IR) studies identified a general cluster of genes up-regulated due to different infectious agents (Dawson et al. 2013). This cluster was also up-regulated after a PRRSv infection in both alveolar macrophages and lymph nodes, albeit at a slower pace when comparing to *Salmonella* spp. infection or stimulation with LPS (Dawson et al. 2013). Using all available porcine IR microarray data, including many array platforms, Badaoui et al. (2013) performed a meta-analysis using the software Pointillist. They compared multiple PRRS microarray studies including many different breeds, tissues and viral strains with many immune response experiments to find PRRS-specific expression responses (Badaoui et al. 2013). Several interferon regulatory transcription factors (IRF1, IRF3, IRF5 and IRF8) were among those found to respond to immune stimulation only in PRRS-specific experiments. In an extension of the WUR-specific transcriptomic analysis by Koltes et al. (2015), Schroyen et al. (2016) looked at the whole transcriptome in order to find differences in pathways between the different genotypes and found pathway differences as a result of the inability of the truncated GBP5 protein in susceptible pigs to restrain viral entry and replication as fast as the intact GBP5 protein in the more resistant pigs.

More recently, Loving et al. (in preparation) performed RNA-seq studies on thymus from non-infected animals and animals infected with different PRRSv strains to investigate thymic atrophy during the infection and how this is reflected in the thymic transcriptome. Thymic samples were collected from four groups of ±5 animals per group (non-infected animals and animals infected with a mild, moderate and severe strain) at 4 and 10 dpi. The number of up-regulated and down-regulated genes between the non-infected and infected animals increased with severity of strain. The transcriptome of the animals infected with the mild or moderate strain showed an inflammatory response at 4 dpi but the infection was resolved by 10 dpi, whereas for the most virulent strain, inflammation was still present at 10 dpi. The most severe PRRSv strain also caused the largest impact on thymic atrophy due to apoptosis, so that the amount and types of cells should be taken into account to fully understand the data. This experiment is therefore a further illustration of the impact of cell counts, as described above for blood transcriptomics.

Since miRNAs play an important role in influencing gene expression levels in a post-transcriptional manner, especially during an immune response (Contreras and Rao 2012), the miRNAome has also been examined with regard to PRRS infection. Several miRNAs are differentially expressed between infected and non-infected animals (Hicks et al. 2013), and there are responses unique to different PRRSv strains (Cong et al. 2014) or within different pig breeds (Li et al. 2015a). Interestingly, in two studies published this year, several miRNAs that were previously identified as influencing innate immunity or have antiviral functions were tested for their ability to reduce PRRSv in infected alveolar macrophages or MARC-145 cells. Jia et al. (2015) transfected MARC-145 cells with 10 miRNAs and at 24 h after transfection infected them with PRRSv at multiplicity of infection (MOI) of 0.1. Compared with the other miRNAs, a fivefold reduction of the viral titer was shown at 72 hours post-inoculation (hpi) when the cells were transfected with miR-26a. PRRSv also induced miR-26a expression in a dose-dependent manner. Li et al. (2015b) looked at 15 miRNAs in both alveolar macrophages or MARC-145 cells and found similar results at a MOI of 0.01 with a 25% and 50% reduction of viral titer at 24 and 48 hpi, respectively, when cells were transfected with miR-26a. Both groups used a luciferase reporter analysis to show that the overexpression of miR-26a affects PRRSv infection, not by attacking the PRRS genome directly but by up-regulation of the innate antiviral response and activation of type I interferon and interferon-induced genes (Jia et al. 2015; Li et al. 2015b).

4.3 Initial Proteomics Approaches to Understanding Host Response to PRRSv

Using iTRAQ labeling, Lu et al. (2012) examined the proteome in PAMs during PRRSv infection. A total of 160 proteins were differentially expressed between uninfected animals and infected animals for at least one time point from 12 up to 48 h post-inoculation of the cells with the virus (Lu et al. 2012). Among them were proteins involved in cytoskeleton networks and cell–cell communication, which is not surprising since viruses can hijack or interact with the host cytoskeletal transport machinery (Dohner and Sodeik 2005). This result was recently confirmed (and extended), as an RNA-seq analysis of blood also saw differences in network connections of genes involved in cytoskeleton rearrangement between susceptible and more resistant pigs (Schroyen et al. 2016). Other DE proteins found were involved in the oxidation-reduction system, RNA-binding proteins or heat shock proteins, which was also reported in other proteomics studies performed on porcine alveolar macrophages (PAMs) or lungs after PRRSv infection (Lu et al. 2012; Xiao et al. 2010; Zhang et al. 2009). However, the question remains how specific these proteins are up-regulated due to the PRRS virus, in contrast with the response to other viruses.

In order to find biomarker proteins in serum to detect early-onset PRRSv infection, Genini et al. (2012) used surface-enhanced laser desorption ionization time of flight mass spectrometry (SELDI-TOF MS). At the day of serum collection, no clinical signs were noted, and none of the piglets were treated. Genini et al. (2012) were able to find a set of 14 discriminatory proteins that could assign pigs to PRRSv-negative and PRRSv-positive groups with high accuracy. They used a dataset of 50 piglet serum samples (from 25 PRRS positive and 25 PRRS negative pigs) to discover these proteins and validated this set in an additional 70 serum samples from 35 PRRS positive and 35 PRRS negative pigs (Genini et al. 2012). We compared these 14 proteins with mRNA information from transcriptomic studies examining host response to PRRSv and some of these proteins could be linked directly to DE or differentially wired (DW) genes, while others belonged to families of genes that were DE in those studies. One of the 14 proteins was the S100 calcium-binding protein A10 (S100A10) and Miller et al. (2012) identified three family members (*S100A8*, *S100A9* and *S100A12*) among the top 10 up-regulated genes after PRRSv infection. This DE occurred at the mRNA level in trachea–bronchial lymph nodes when animals infected with PRRS were compared to non-infected animals (Miller et al. 2012). Other interesting proteins among those 14 biomarkers were proteasome activator family member 28 beta, ubiquitin and vacuolar protein sorting 29 (vps29). Interestingly, in Schroyen et al. (2016), proteasome activator family member 28 beta (*PSME2*) and ubiquitin protein ligase E3A (*UBE3A*) were DW between susceptible and more resistant animals. Furthermore, *VPS41* had a high phenotypic impact factor, which meant that it was DE between the susceptible and more resistant animals and at the same time highly expressed (Schroyen et al. 2016).

Recently, Luo et al. (2014) were the first to examine the PRRSv host response phosphoproteome, a large-scale study of protein phosphorylation levels in PAMs, using a TiO_2-based enrichment method combined with liquid chromatography tandem mass spectrometry (LC-MS/MS). The phosphorylation level of over 200 proteins was altered at both 12 and 36 h post-infection (Luo et al. 2014). Pathway analysis revealed that several signal transduction pathways such as MAPK, NF-kB and PI3K-AKT signaling pathways were significantly activated after infection. It has been reported that the PI3K-Akt signal transduction pathway is involved in PRRSv entry (Ni et al. 2015; Zhu et al. 2013).

4.4 Mathematical models to help in the integration of PRRS data

A systems biology approach to understand the host response to PRRS would integrate these genomic, transcriptomic and proteomic results. Alternatively, mathematical host–pathogen interaction models could integrate these diverse empirical findings and contribute to enhancing our understanding of the immune responses even further (Doeschl-Wilson 2011). A useful example of mathematical modeling of host–PRRS interactions has been provided by Doeschl-Wilson and Galina-Pantoja (2010). Such modeling approaches start off as basic host–pathogen models describing the interaction between virus and host macrophages without host immune response, and increase complexity gradually by adding innate, humoral and cellular immune responses (Doeschl-Wilson and Galina-Pantoja 2010). Besides giving better insights, such models can also point towards missing system components and open up to further experimental investigations.

Doeschl-Wilson et al. (2012) applied the dynamical systems theory on individuals after a PRRSv infection. They could distinguish nine different performances versus pathogen burden trajectories in pigs infected with the same dose of PRRSv. They propose to use these trajectories as reliable categorical tolerance phenotypes in subsequent genetic studies (Doeschl-Wilson et al. 2012). While inspecting the viremia patterns in the blood over a time period from 0 dpi to 42 dpi, another categorical distinction emerged: cleared, persistent and rebound phenotypes. Islam et al. (2013) used Wood's curves to fit these blood viremia patterns and linked the analysis of neutralizing antibody (nAb) to these patterns (Islam et al. 2013). In the pigs that were classified as cleared, a narrow nAb response was noted, showing an efficient immune response by which the virus used in the infectious dose is rapidly cleared. Pigs that were persistently viremic over the 42-day period displayed a broad nAb spectrum, indicating a more inefficient antibody response to the original strain as well as potentially a more diverse response due to new viral quasi-species that arise from within the inoculum via selection pressure from the host immune response. It would be of great interest to link these different types of immune response to transcriptomic and/or proteomic data and identify markers for successful adaptive immunity to PRRSv.

4.5 Systems Biology on PRRS

In some of the studies described above, some form of "systems-wide analyses" was utilized. When RNA-seq is performed to examine expression differences between animals with a different genotype for an SNP marker related to viral load and weight gain, transcriptomics meets genomics (e.g. Koltes et al. 2015, Schroyen et al. 2016). When the gene ontology analysis of genes in the vicinity of genetic markers associated with response traits elucidates differentially expressed pathways between susceptible and more resistant animals, genomics meets transcriptomics (e.g. Waide et al. submitted). When the genes encoding proteins found with proteomics are also identified by using RNA-seq analyses, or when altered expression of phosphorylation levels are found in proteins of a specific pathway, whose genes are up-regulated or down-regulated in microarray or RNA-seq experiments, proteomics meets transcriptomics (Genini et al. 2012; Lu et al. 2012; Luo et al. 2014; Miller et al. 2012; Schroyen et al. 2016).

To integrate the data from our whole blood microarray experiment described in Schroyen et al. (2015) with knowledge on protein interaction data, we re-analyzed the genes found in the immune-related module and performed a protein–protein interaction (PPI) analysis on these genes using NetworkAnalyst (Xia et al. 2014). By firstly annotating the genes in this module, it could be seen that the cluster is enriched for interesting annotations, including cytokines, chemokines, interferon type I stimulated genes, apoptotic genes and genes involved in complement pathways. Because all genes were allocated to the same co-expression module, their mRNA expression pattern from animal to animal was similar. By using NetworkAnalyst, knowledge about existing (human) protein–protein interactions is added on top of the mRNA information. We determined the largest zero-order interaction network between proteins encoded by the 506 genes in the immune-related module and found a set of 33 proteins, of which the topology is shown in Fig. 2a. In Schroyen et al. (2015), components of this protein network were identified, namely the inflammasome gene *NLRP3*, which is known to activate *CASP1* and in turn leads to the activation of *IL1B* and *IL18*. However, with the PPI analysis, other connections become clear. For instance, *TXNIP* was found DE in PRRSv-infected lungs and bronchial lymph nodes (Bates et al. 2008) and its protein interacts with the NLRP3 protein. The pathogen-recognition RIG1 receptor or DDX58 interacts with CASP1, which in turn is linked to the interferon-stimulated IFIT3. The anti-apoptosis BCL2 family member MCL1 is linked to CASP1 through CASP3. To further explore this PPI network, the genes in this PPI network that exhibit up-regulation or down-regulation after 4 dpi compared to 0 dpi is shown in Fig. 2b. Because the animals in this microarray experiment had been genotyped for the WUR SNP described earlier by Boddicker et al. (2012) as a marker for

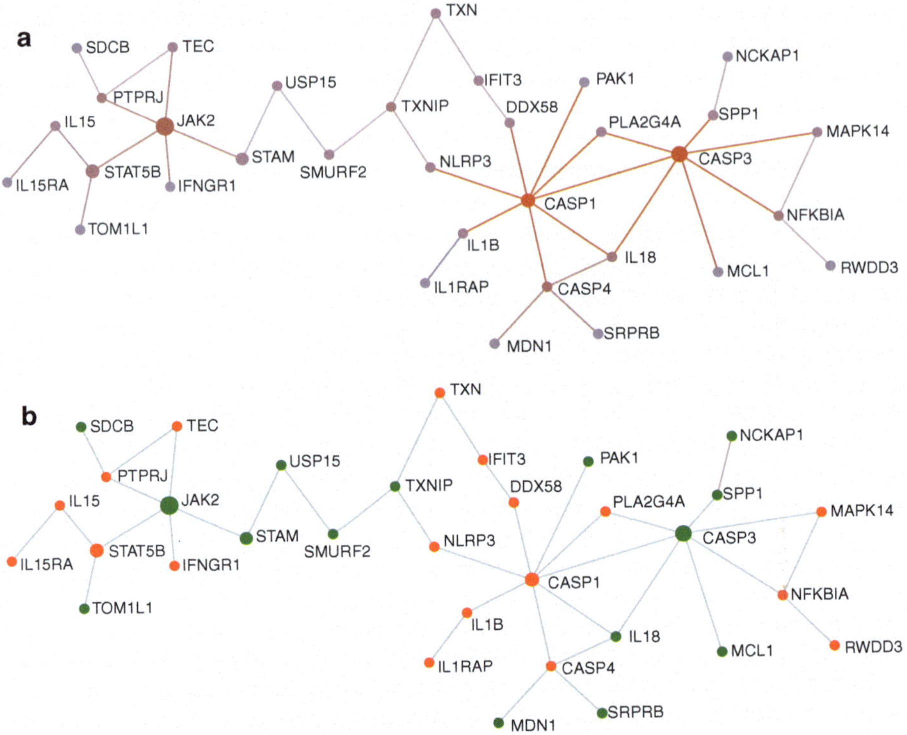

Fig. 2 NetworkAnalyst protein–protein interactions (PPI) on immune-related module found in the whole blood PRRS microarray study of Schroyen et al. 2015. (**a**) Nodes are colored according to connectivity; more red means more connections. (**b**) Nodes are colored according to up-regulation (*red*) and down-regulation (*green*) of genes at 4 dpi compared to 0 dpi after PRRS infection

susceptibility, the immune-related module was also correlated with WUR genotype. Looking at this reduced protein network, the substantial differences in expression pattern between more and less susceptible animals is very clear (Fig. 3a, b). These multi-omics analyses can help us better understand biological processes such as immune responses and they can be used to confirm or reject hypotheses made after performing a single-omics study. In any case, more information can be gained, often at low cost. As with the PPI example, the introduction of a protein network on top of transcriptomic data displayed a distinct small subset of 33 correlated genes that was evidently different between WUR genotypes animals and was not visible when looking at the micro-array dataset alone.

Example 2: Systems Biology in *Salmonella* Studies in Pig

Another important pathogen in the swine industry is *Salmonella*. It is a foodborne pathogen hazardous for human consumption, causing severe gastroenteritis and deaths worldwide. In the USA alone, costs for human salmonellosis are estimated

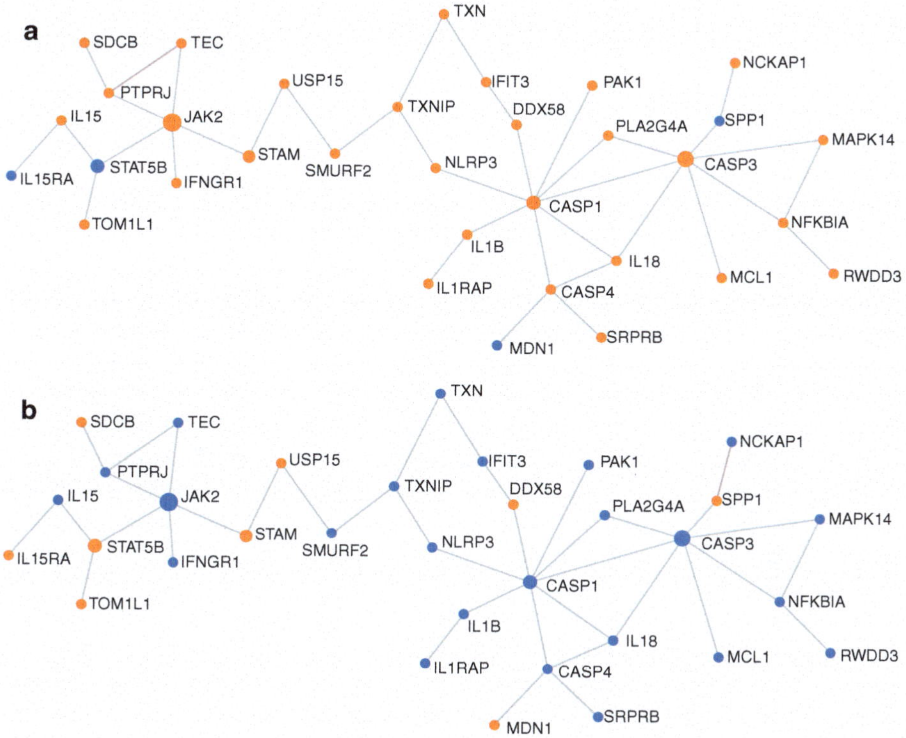

Fig. 3 NetworkAnalyst protein–protein interactions (PPI) on immune-related module found in the whole blood PRRS microarray study of Schroyen et al. 2015, split between animals with the different WUR marker, predicting for susceptibility to PRRS found by Boddicker et al. 2012. (**a**) Nodes are colored according to positive average expression (*orange*) and negative average expression (*blue*) after LIMMA normalization of microarray data of genes (Schroyen et al. 2015) in more susceptible animals at 4 dpi. (**b**) Nodes are colored according to positive average expression (*orange*) and negative average expression (*blue*) after normalization of genes in less susceptible animals at 4 dpi

at more than $2.4 billion annually. Human salmonellosis can often be linked to an animal source such as poultry, eggs, pork, beef and dairy cattle (Callaway et al. 2008). Other than affecting human health, *Salmonella* spp. also infect and/or multiply in almost all known vertebrates, from reptiles to birds and mammals (Edwards et al. 2002), and clinical and subclinical salmonellosis in pigs has been estimated to contribute to substantial economic losses to the swine industry (Haley et al. 2012).

4.6 Network-Based Analysis of *Salmonella* in Pigs

Probably the two most examined *Salmonella* serovars concerning pig gene expression regulation are *S. enterica* serovar Typhimurium (ST) and *S. enterica* serovar

Choleraesuis (SC). ST causes enterocolitis in a wide variety of vertebrates, while ST is host-adapted and predominantly affects swine (Edwards et al. 2002). In pigs, SC was the most common serovar from 1986 to 1995, but in the mid-1990s, it was replaced by ST (Foley et al. 2008). Recently, several transcriptomic studies were performed to determine differences in whole blood causing variation in outcome between low (LS) and persistently shedding (PS) pigs after inoculation with ST (Huang et al. 2011; Knetter et al. 2015; Uthe et al. 2009, 2011). In order to find biomarkers that could distinguish between LS and PS animals before infection, Kommadath et al. (2014) performed a network-based analysis. Using recently acquired RNA-seq data of blood from ST-infected pigs and WGCNA, they found day 0 modules that contained genes annotated for innate defense against bacteria—or Salmonella in particular—and that had distinct expression patterns in LS versus PS animals, with the mean expression levels higher in the LS than PS animals. Examining the connectivity of the genes revealed that connections to hub genes within these modules were significantly stronger in LS than PS animals, which could be an indication of a more tightly regulated transcriptional response of the genes in these modules in the LS animals (Kommadath et al. 2014), and supports the hypothesis that LS animals are better prepared for an infection and quicker to respond.

miRNA-seq was performed on whole blood samples of the same set of LS and PS animals and together with the mRNA-seq data used by Kommadath et al. (2014), a potential involvement of miRNAs was examined (Bao et al. 2015). In both LS and PS pigs, miR-214 and miR-331-3p were associated with ST infection. Targets for miR-214 were predicted to be *SLC11A1* and *LILR*-like. The expression of the mRNA for these two genes increased at 2 dpi, while the expression of miR-214 expression decreased. Both these genes are involved in immune response, but no role for miRNAs to control them has yet been described. *VAV2* plays a role in the entry process of several pathogenic microbes. It is a target gene for miR-331-3p and had a lower expression after infection, which could be the result of an observed increase in miR-331-3p expression. Results were of a similar magnitude in both LS and PS animals. For comparisons between LS and PS, no miRNAs were DE at 0 dpi, and only three were DE at 2 dpi. Bao et al. (2014), as described earlier, reported a more tightly rewired network after *Salmonella* infection, and it would be interesting to look at DW between LS and PS animals of target mRNAs at 0 dpi.

4.7 Salmonella and the Microbiome

The pig microbiome has been the subject of many immune-related studies and gut microbiota are widely recognized to play a crucial role in animal health and well-being (Kim and Isaacson 2015). Bearson et al. (2013) compared the microbiome in non-infected (NI), LS and PS animals at days 0, and 2, 7 and at 21 dpi. At 0 dpi, significant differences in microbial community structure were seen between LS and PS animals; however, these two groups were both not significantly different from the NI group. At 2 and 7 dpi, there was no difference in the microbiome between the

LS group and the NI animals, but a clear difference was shown between PS and the other two groups of animals. At 21 dpi, these differences between LS and PS groups were gone; however, microbiota profiles for both LS and PS were significantly different from the NI group at 21 dpi, suggesting a Salmonella-induced alteration in microbiota regardless of shedding status (Bearson et al. 2013).

With regard to screening for biomarkers for resistance/tolerance versus susceptibility before infection, DNA sequence analysis of day 0 microbiota samples in this study revealed an enriched presence of *Ruminococcaceae* in the LS animals (Bearson et al. 2013). This positive effect of *Ruminococcaceae* on resistance/tolerance is described in several studies focusing on intestinal microbiota compositions with regard to diarrhea, whether caused by *Salmonella* spp. or not (Pop et al. 2014; Suchodolski et al. 2012; Videnska et al. 2013). Members of this microbial family produce short-chain fatty acids (SCFA) with acetate, butyrate and propionate being the major SCFA produced in the colon. Specifically, butyrate can be influential in gut health due to its anti-inflammatory properties and its capacity to strengthen the colonic barrier and reduce the intestinal epithelial permeability (Hamer et al. 2008). In a study on gut microbiota in children with eczema, a negative association was reported between *Ruminococcaceae* and TLR2-induced IL6 and TNFα levels (West et al. 2015). Earlier, Huang et al. (2011) found that only in PS pigs, TNFα RNA in blood was elevated after 2 dpi ST infection (Huang et al. 2011). One interpretation of these results is that PS animals, with less *Ruminococcaceae* in their intestine compared to LS animals, do elevate the TNFα pathway, whereas in LS animals this is not the case. Certainly, more research is required to ascertain the generality of these proposed relationships.

5 Current Challenges and Future Directions

In the pig, there are only a handful of examples of studies approaching a systems biology analysis described thus far, but the merit of such research is becoming more and more apparent. Immunology is a highly relevant research domain for a systems-level approach because of the multitude of tissues, cells, proteins or genes interacting with one another when facing a disease challenge, with such interactions occurring at multiple scales of time. Currently, data created and analyzed by different labs and different experiments are hard to integrate in a powerful way due to different breeds used, different time points examined, and different protocols followed. To make a systems biology approach easier, consortia led by a complementary set of laboratories or institutions are being established (Benoist et al. 2012). Genetics research is far more active in consortium science, since it is easier to identify, map or sequence genes by several groups than it is to examine a complex immunological research question (Benoist et al. 2012). For pig, the PiGMaP consortium (Archibald et al. 1995) and the Swine Genome Sequencing Consortium (Schook et al. 2005) were the first consortia established. For pig diseases, and specifically to examine PRRS virus infections in pigs, the PRRS Host Genetics Consortium (PHGC) was founded (Lunney et al. 2011). Some of the research described above is part of this consortium (Boddicker et al. 2012, 2013, 2014; Koltes et al. 2015; Schroyen et al. 2015, 2016; Waide et al. submitted), and the genetic and immunological insights gained strongly demonstrate the value of collaborative efforts that increase the power of such challenge experiments.

A substantial advantage of these consortia is that variation is reduced by shared and standardized protocols and procedures, as is described for the Human Encyclopedia of DNA Elements (ENCODE) project (ENCODE project consortium 2011). By using standards, data quality is assured, data utility can be extended and data comparison and thus the establishment of a systems biology approach, has become easier. The ENCODE project has been expanded from humans to classical model species and recently the Functional Annotation of Animal Genomes (FAANG) consortium for domesticated animal species was launched (The FAANG consortium et al. 2015). As a start, this consortium will focus on chicken, pig, cattle, horse, goat and sheep, species with a high-quality reference genome and often a plentitude of (ancestor's) phenotypic data already stored (The FAANG consortium et al. 2015). Cells and tissues relevant to pig health, including blood cells and liver, are being collected on healthy pigs in the FAANG project (www.faang.org). In addition, several groups have pathogen challenge projects that will provide data relevant to a deeper understanding of the porcine immune response and the parts of the genome that are responsible for these responses. Thus, the FAANG project will accelerate our ability to apply systems biology tools to improving pig health in the future.

References

Abbas AR, Baldwin D, Ma Y, Ouyang W, Gurney A, Martin F, Fong S, van Lookeren Campagne M, Godowski P, Williams PM, Chan AC, Clark HF (2005) Immune response in silico (IRIS): immune-specific genes identified from a compendium of microarray expression data. Genes Immun 6:319–331

Ait-Ali T, Wilson AD, Carre W, Westcott DG, Frossard JP, Mellencamp MA, Mouzaki D, Matika O, Waddington D, Drew TW, Bishop SC, Archibald AL (2011) Host inhibits replication of European porcine reproductive and respiratory syndrome virus in macrophages by altering differential regulation of type-I interferon transcriptional response. Immunogenetics 63:437–448

Arakawa A, Okumura N, Taniguchi M, Hayashi T, Hirose K, Fukawa K, Ito T, Matsumoto T, Uenishi H, Mikawa S (2015) Genome-wide association QTL mapping for teat number in a purebred population of Duroc pigs. Anim Genet 46(5):571–575

Archibald AL, Haley CS, Brown JF, Couperwhite S, McQueen HA, Nicholson D, Coppieters W, Van de Weghe A, Stratil A, Wintero AK et al (1995) The PiGMaP consortium linkage map of the pig (Sus scrofa). Mamm Genome 6:157–175

Badaoui B, Tuggle CK, Hu Z, Reecy JM, Ait-Ali T, Anselmo A, Botti S (2013) Pig immune response to general stimulus and to porcine reproductive and respiratory syndrome virus infection: a meta-analysis approach. BMC Genomics 14:220

Badaoui B, Rutigliano T, Anselmo A, Vanhee M, Nauwynck H, Giuffra E, Botti S (2014) RNA-sequence analysis of primary alveolar macrophages after in vitro infection with porcine reproductive and respiratory syndrome virus strains of differing virulence. PLoS One 9, e91918

Bao H, Kommadath A, Plastow GS, Tuggle CK, le Guan L, Stothard P (2014) MicroRNA buffering and altered variance of gene expression in response to Salmonella infection. PLoS One 9, e94352

Bao H, Kommadath A, Liang G, Sun X, Arantes AS, Tuggle CK, Bearson SMD, Plastow GS, Stothard P, Guan LL (2015) Genome-Wide whole blood microRNAome and transcriptome analyses revealed miRNA-mRNA regulated host response to foodborne pathogen *Salmonella* infection in swine. Sci Rep 5:12620

Bates JS, Petry DB, Eudy J, Bough L, Johnson RK (2008) Differential expression in lung and bronchial lymph node of pigs with high and low responses to infection with porcine reproductive and respiratory syndrome virus. J Anim Sci 86:3279–3289

Bearson SM, Allen HK, Bearson BL, Looft T, Brunelle BW, Kich JD, Tuggle CK, Bayles DO, Alt D, Levine UY, Stanton TB (2013) Profiling the gastrointestinal microbiota in response to Salmonella: low versus high Salmonella shedding in the natural porcine host. Infect Genet Evol 16:330–340

Bebek G, Koyuturk M, Price ND, Chance MR (2012) Network biology methods integrating biological data for translational science. Brief Bioinform 13:446–459

Benoist C, Lanier L, Merad M, Mathis D (2012) Consortium biology in immunology: the perspective from the Immunological Genome Project. Nat Rev Immunol 12:734–740

Boddicker N, Waide EH, Rowland RR, Lunney JK, Garrick DJ, Reecy JM, Dekkers JC (2012) Evidence for a major QTL associated with host response to porcine reproductive and respiratory syndrome virus challenge. J Anim Sci 90:1733–1746

Boddicker NJ, Garrick DJ, Rowland RR, Lunney JK, Reecy JM, Dekkers JC (2013) Validation and further characterization of a major quantitative trait locus associated with host response to experimental infection with porcine reproductive and respiratory syndrome virus. Anim Genet 45:48–58

Boddicker NJ, Bjorkquist A, Rowland RR, Lunney JK, Reecy JM, Dekkers JC (2014) Genome-wide association and genomic prediction for host response to porcine reproductive and respiratory syndrome virus infection. Genet Sel Evol 46:18

Callaway TR, Edrington TS, Anderson RC, Byrd JA, Nisbet DJ (2008) Gastrointestinal microbial ecology and the safety of our food supply as related to Salmonella. J Anim Sci 86:E163–E172

Carbon S, Ireland A, Mungall CJ, Shu S, Marshall B, Lewis S (2009) AmiGO: online access to ontology and annotation data. Bioinformatics 25:288–289

Chaussabel D (2015) Assessment of immune status using blood transcriptomics and potential implications for global health. Semin Immunol 27:58–66

Chaussabel D, Pascual V, Banchereau J (2010) Assessing the human immune system through blood transcriptomics. BMC Biol 8:84

Chomwisarutkun K, Murani E, Brunner R, Ponsuksili S, Wimmers K (2013) QTL region-specific microarrays reveal differential expression of positional candidate genes of signaling pathways associated with the liability for the inverted teat defect. Anim Genet 44:139–148

Clapperton M, Diack AB, Matika O, Glass EJ, Gladney CD, Mellencamp MA, Hoste A, Bishop SC (2009) Traits associated with innate and adaptive immunity in pigs: heritability and associations with performance under different health status conditions. Genet Sel Evol 41:54

Cong P, Xiao S, Chen Y, Wang L, Gao J, Li M, He Z, Guo Y, Zhao G, Zhang X, Chen L, Mo D, Liu X (2014) Integrated miRNA and mRNA transcriptomes of porcine alveolar macrophages (PAM cells) identifies strain-specific miRNA molecular signatures associated with H-PRRSV and N-PRRSV infection. Mol Biol Rep 41:5863–5875

Contreras J, Rao DS (2012) MicroRNAs in inflammation and immune responses. Leukemia 26:404–413

da Huang W, Sherman BT, Lempicki RA (2009) Systematic and integrative analysis of large gene lists using DAVID bioinformatics resources. Nat Protoc 4:44–57

Dawson HD, Loveland JE, Pascal G, Gilbert JG, Uenishi H, Mann KM, Sang Y, Zhang J, Carvalho-Silva D, Hunt T, Hardy M, Hu Z, Zhao SH, Anselmo A, Shinkai H, Chen C, Badaoui B, Berman D, Amid C, Kay M, Lloyd D, Snow C, Morozumi T, Cheng RP, Bystrom M, Kapetanovic R, Schwartz JC, Kataria R, Astley M, Fritz E, Steward C, Thomas M, Wilming L, Toki D, Archibald AL, Bed'Hom B, Beraldi D, Huang TH, Ait-Ali T, Blecha F, Botti S, Freeman TC, Giuffra E, Hume DA, Lunney JK, Murtaugh MP, Reecy JM, Harrow JL, Rogel-Gaillard C, Tuggle CK (2013) Structural and functional annotation of the porcine immunome. BMC Genomics 14:332

Doeschl-Wilson AB (2011) The role of mathematical models of host-pathogen interactions for livestock health and production – a review. Animal 5:895–910
Doeschl-Wilson A, Galina-Pantoja L (2010) Using Mathematical Models to Gain Insight into Host-Pathogen Interaction in Mammals: Porcine Reproductive and Respiratory Syndrome. In: Barton AW (ed) Host-Pathogen Interactions: Genetics, Immunology, Physiology. Nova Science Publishers Inc, New York, p 109–131
Doeschl-Wilson AB, Bishop SC, Kyriazakis I, Villanueva B (2012) Novel methods for quantifying individual host response to infectious pathogens for genetic analyses. Front Genet 3:266
Dohner K, Sodeik B (2005) The role of the cytoskeleton during viral infection. Curr Top Microbiol Immunol 285:67–108
Eden E, Navon R, Steinfeld I, Lipson D, Yakhini Z (2009) GOrilla: a tool for discovery and visualization of enriched GO terms in ranked gene lists. BMC Bioinformatics 10:48
Edwards RA, Olsen GJ, Maloy SR (2002) Comparative genomics of closely related salmonellae. Trends Microbiol 10:94–99
ENCODE Project Consortium (2011) A user's guide to the encyclopedia of DNA elements (ENCODE). PLoS Biol 9, e1001046
Endale Ahanda ML, Fritz ER, Estelle J, Hu ZL, Madsen O, Groenen MA, Beraldi D, Kapetanovic R, Hume DA, Rowland RR, Lunney JK, Rogel-Gaillard C, Reecy JM, Giuffra E (2012) Prediction of altered 3'- UTR miRNA-binding sites from RNA-Seq data: the swine leukocyte antigen complex (SLA) as a model region. PLoS One 7, e48607
Ernst CW, Steibel JP (2013) Molecular advances in QTL discovery and application in pig breeding. Trends Genet 29:215–224
Flori L, Gao Y, Laloe D, Lemonnier G, Leplat JJ, Teillaud A, Cossalter AM, Laffitte J, Pinton P, de Vaureix C, Bouffaud M, Mercat MJ, Lefevre F, Oswald IP, Bidanel JP, Rogel-Gaillard C (2011) Immunity traits in pigs: substantial genetic variation and limited covariation. PLoS One 6, e22717
Foley SL, Lynne AM, Nayak R (2008) Salmonella challenges: prevalence in swine and poultry and potential pathogenicity of such isolates. J Anim Sci 86:E149–E162
Fowler KE, Pong-Wong R, Bauer J, Clemente EJ, Reitter CP, Affara NA, Waite S, Walling GA, Griffin DK (2013) Genome wide analysis reveals single nucleotide polymorphisms associated with fatness and putative novel copy number variants in three pig breeds. BMC Genomics 14:784
Freeman TC, Goldovsky L, Brosch M, van Dongen S, Maziere P, Grocock RJ, Freilich S, Thornton J, Enright AJ (2007) Construction, visualisation, and clustering of transcription networks from microarray expression data. PLoS Comput Biol 3:2032–2042
Freeman TC, Ivens A, Baillie JK, Beraldi D, Barnett MW, Dorward D, Downing A, Fairbairn L, Kapetanovic R, Raza S, Tomoiu A, Alberio R, Wu C, Su AI, Summers KM, Tuggle CK, Archibald AL, Hume DA (2012) A gene expression atlas of the domestic pig. BMC Biol 10:90
Galina-Pantoja L, Torremorell M, Deeb N, geiger B, Gladney C, Mellencamp MA (2006) DNA markers associated with reproductive traits during PRRSv infection. In: Proceedings of the 19th IVIS IPVS Congress Denmark
Genini S, Delputte PL, Malinverni R, Cecere M, Stella A, Nauwynck HJ, Giuffra E (2008) Genome-wide transcriptional response of primary alveolar macrophages following infection with porcine reproductive and respiratory syndrome virus. J Gen Virol 89:2550–2564
Genini S, Paternoster T, Costa A, Botti S, Luini MV, Caprera A, Giuffra E (2012) Identification of serum proteomic biomarkers for early porcine reproductive and respiratory syndrome (PRRS) infection. Proc Natl Acad Sci U S A 10:48
Haley CA, Dargatz DA, Bush EJ, Erdman MM, Fedorka-Cray PJ (2012) Salmonella prevalence and antimicrobial susceptibility from the National Animal Health Monitoring System Swine 2000 and 2006 studies. J Food Prot 75:428–436
Hamer HM, Jonkers D, Venema K, Vanhoutvin S, Troost FJ, Brummer RJ (2008) Review article: the role of butyrate on colonic function. Aliment Pharmacol Ther 27:104–119
Hicks J, Yoo D, Liu HC (2013) Characterization of the microRNAome in porcine reproductive and respiratory syndrome virus infected macrophages. PLoS One 8, e82054

Hollung K, Timperio AM, Olivan M, Kemp C, Coto-Montes A, Sierra V, Zolla L (2014) Systems biology: a new tool for farm animal science. Curr Protein Pept Sci 15:100–117

Holtkamp DJ, Kliebenstein JB, Neumann EJ, Zimmerman JJ, Rotto HF, Yoder TK, Wang C, Yeske PE, Mowrer CL, Haley CA (2013) Assessment of the economic impact of porcine reproductive and respiratory syndrome virus on United States pork producers. J Swine Health Prod 21:72–84

Huang TH, Uthe JJ, Bearson SM, Demirkale CY, Nettleton D, Knetter S, Christian C, Ramer-Tait AE, Wannemuehler MJ, Tuggle CK (2011) Distinct peripheral blood RNA responses to Salmonella in pigs differing in Salmonella shedding levels: intersection of IFNG, TLR and miRNA pathways. PLoS One 6, e28768

Hulst M, Loeffen W, Weesendorp E (2013) Pathway analysis in blood cells of pigs infected with classical swine fever virus: comparison of pigs that develop a chronic form of infection or recover. Arch Virol 158:325–339

Islam ZU, Bishop SC, Savill NJ, Rowland RR, Lunney JK, Trible B, Doeschl-Wilson AB (2013) Quantitative analysis of porcine reproductive and respiratory syndrome (PRRS) viremia profiles from experimental infection: a statistical modelling approach. PLoS One 8, e83567

Jegou M, Gondret F, Vincent A, Trefeu C, Gilbert H, Louveau I (2016) Whole blood transcriptomics is relevant to identify molecular changes in response to genetic selection for feed efficiency and nutritional status in the pig. PLoS One 11, e0146550

Jia X, Bi Y, Li J, Xie Q, Yang H, Liu W (2015) Cellular microRNA miR-26a suppresses replication of porcine reproductive and respiratory syndrome virus by activating innate antiviral immunity. Sci Rep 5:10651

Kadarmideen HN, Watson-Haigh NS (2012) Building gene co-expression networks using transcriptomics data for systems biology investigations: Comparison of methods using microarray data. Bioinformation 8:855–861

Kanehisa M, Goto S, Sato Y, Kawashima M, Furumichi M, Tanabe M (2014) Data, information, knowledge and principle: back to metabolism in KEGG. Nucleic Acids Res 42:D199–D205

Kapetanovic R, Fairbairn L, Downing A, Beraldi D, Sester DP, Freeman TC, Tuggle CK, Archibald AL, Hume DA (2013) The impact of breed and tissue compartment on the response of pig macrophages to lipopolysaccharide. BMC Genomics 14:581

Kidd BA, Peters LA, Schadt EE, Dudley JT (2014) Unifying immunology with informatics and multiscale biology. Nat Immunol 15:118–127

Kim HB, Isaacson RE (2015) The pig gut microbial diversity: understanding the pig gut microbial ecology through the next generation high throughput sequencing. Vet Microbiol 177:242–251

Knetter SM, Bearson SM, Huang TH, Kurkiewicz D, Schroyen M, Nettleton D, Berman D, Cohen V, Lunney JK, Ramer-Tait AE, Wannemuehler MJ, Tuggle CK (2015) Salmonella enterica serovar Typhimurium-infected pigs with different shedding levels exhibit distinct clinical, peripheral cytokine and transcriptomic immune response phenotypes. Innate Immun 21:227–241

Koesterke L, Milfeld K, Vaughn M, Stanzione D, Koltes J, Weeks N, Reecy J (2013) Optimizing the PCIT algorithm on stampede's Xeon and Xeon Phi processors for faster discovery of biological networks. In: Proceedings of the conference on extreme science and engineering discovery environment: gateway to discovery

Koesterke L, Koltes J, Weeks N, Milfeld K, Vaughn M, Reecy J, Stanzione D (2014) Discovery of biological networks using an optimized partial correlation coefficient with information theory algorthim on Stampede's Xeon and Xeons Phi processors. Concurrency Comput Pract Exp 26:2178–2190

Kogelman LJ, Cirera S, Zhernakova DV, Fredholm M, Franke L, Kadarmideen HN (2014) Identification of co-expression gene networks, regulatory genes and pathways for obesity based on adipose tissue RNA Sequencing in a porcine model. BMC Med Genomics 7:57

Koltes JE, Fritz-Waters E, Eisley CJ, Choi I, Bao H, Kommadath A, Serao NV, Boddicker NJ, Abrams SM, Schroyen M, Loyd H, Tuggle CK, Plastow GS, Guan L, Stothard P, Lunney JK, Liu P, Carpenter S, Rowland RR, Dekkers JC, Reecy JM (2015) Identification of a putative quantitative trait nucleotide in guanylate binding protein 5 for host response to PRRS virus infection. BMC Genomics 16:412

Kommadath A, Bao H, Arantes AS, Plastow GS, Tuggle CK, Bearson SM, le Guan L, Stothard P (2014) Gene co-expression network analysis identifies porcine genes associated with variation in Salmonella shedding. BMC Genomics 15:452

Langfelder P, Horvath S (2008) WGCNA: an R package for weighted correlation network analysis. BMC Bioinformatics 9:559

Li J, Chen Z, Zhao J, Fang L, Fang R, Xiao J, Chen X, Zhou A, Zhang Y, Ren L, Hu X, Zhao Y, Zhang S, Li N (2015a) Difference in microRNA expression and editing profile of lung tissues from different pig breeds related to immune responses to HP-PRRSV. Sci Rep 5:9549

Li L, Wei Z, Zhou Y, Gao F, Jiang Y, Yu L, Zheng H, Tong W, Yang S, Shan T, Liu F, Xia T, Tong G (2015b) Host miR-26a suppresses replication of porcine reproductive and respiratory syndrome virus by upregulating type I interferons. Virus Res 195:86–94

Lu Q, Bai J, Zhang L, Liu J, Jiang Z, Michal JJ, He Q, Jiang P (2012) Two-dimensional liquid chromatography-tandem mass spectrometry coupled with isobaric tags for relative and absolute quantification (iTRAQ) labeling approach revealed first proteome profiles of pulmonary alveolar macrophages infected with porcine reproductive and respiratory syndrome virus. J Proteome Res 11:2890–2903

Lunney JK, Ho CS, Wysocki M, Smith DM (2009) Molecular genetics of the swine major histocompatibility complex, the SLA complex. Dev Comp Immunol 33:362–374

Lunney JK, Steibel JP, Reecy JM, Fritz E, Rothschild MF, Kerrigan M, Trible B, Rowland RR (2011) Probing genetic control of swine responses to PRRSV infection: current progress of the PRRS host genetics consortium. BMC Proc 5(Suppl 4):S30

Luo R, Fang L, Jin H, Wang D, An K, Xu N, Chen H, Xiao S (2014) Label-free quantitative phosphoproteomic analysis reveals differentially regulated proteins and pathway in PRRSV-infected pulmonary alveolar macrophages. J Proteome Res 13:1270–1280

Lynn DJ, Winsor GL, Chan C, Richard N, Laird MR, Barsky A, Gardy JL, Roche FM, Chan TH, Shah N, Lo R, Naseer M, Que J, Yau M, Acab M, Tulpan D, Whiteside MD, Chikatamarla A, Mah B, Munzner T, Hokamp K, Hancock RE, Brinkman FS (2008) InnateDB: facilitating systems-level analyses of the mammalian innate immune response. Mol Syst Biol 4:218

Mach N, Gao Y, Lemonnier G, Lecardonnel J, Oswald IP, Estelle J, Rogel-Gaillard C (2013) The peripheral blood transcriptome reflects variations in immunity traits in swine: towards the identification of biomarkers. BMC Genomics 14:894

McKnite AM, Bundy JW, Moural TW, Tart JK, Johnson TP, Jobman EE, Barnes SY, Qiu JK, Peterson DA, Harris SP, Rothschild MF, Galeota JA, Johnson RK, Kachman SD, Ciobanu DC (2014) Genomic analysis of the differential response to experimental infection with porcine circovirus 2b. Anim Genet 45:205–214

Mellencamp MA, Galina-Pantoja L, Gladney CD, Torremorell M (2008) Improving pig health through genomics: a view from the industry. Dev Biol (Basel) 132:35–41

Mi H, Muruganujan A, Casagrande JT, Thomas PD (2013) Large-scale gene function analysis with the PANTHER classification system. Nat Protoc 8:1551–1566

Miller LC, Fleming D, Arbogast A, Bayles DO, Guo B, Lager KM, Henningson JN, Schlink SN, Yang HC, Faaberg KS, Kehrli ME Jr (2012) Analysis of the swine tracheobronchial lymph node transcriptomic response to infection with a Chinese highly pathogenic strain of porcine reproductive and respiratory syndrome virus. BMC Vet Res 8:208

Mohr S, Liew CC (2007) The peripheral-blood transcriptome: new insights into disease and risk assessment. Trends Mol Med 13:422–432

Ni B, Wen LB, Wang R, Hao HP, Huan CC, Wang X, Huang L, Miao JF, Fan HJ, Mao X (2015) The involvement of FAK-PI3K-AKT-Rac1 pathway in porcine reproductive and respiratory syndrome virus entry. Biochem Biophys Res Commun 458(2):392–398

Ponsuksili S, Du Y, Murani E, Schwerin M, Wimmers K (2012) Elucidating molecular networks that either affect or respond to plasma cortisol concentration in target tissues of liver and muscle. Genetics 192:1109–1122

Pop M, Walker AW, Paulson J, Lindsay B, Antonio M, Hossain MA, Oundo J, Tamboura B, Mai V, Astrovskaya I, Corrada Bravo H, Rance R, Stares M, Levine MM, Panchalingam S, Kotloff K, Ikumapayi UN, Ebruke C, Adeyemi M, Ahmed D, Ahmed F, Alam MT, Amin R, Siddiqui S, Ochieng JB, Ouma E, Juma J, Mailu E, Omore R, Morris JG, Breiman RF, Saha D, Parkhill J, Nataro JP, Stine OC (2014) Diarrhea in young children from low-income countries leads to large-scale alterations in intestinal microbiota composition. Genome Biol 15:R76

Rauw WM (2012) Immune response from a resource allocation perspective. Front Genet 3:267

Reiner G, Dreher F, Drungowski M, Hoeltig D, Bertsch N, Selke M, Willems H, Gerlach GF, Probst I, Tuemmler B, Waldmann KH, Herwig R (2014) Pathway deregulation and expression QTLs in response to Actinobacillus pleuropneumoniae infection in swine. Mamm Genome 25:600–617

Reverter A, Chan EK (2008) Combining partial correlation and an information theory approach to the reversed engineering of gene co-expression networks. Bioinformatics 24:2491–2497

Reverter A, Hudson NJ, Nagaraj SH, Perez-Enciso M, Dalrymple BP (2010) Regulatory impact factors: unraveling the transcriptional regulation of complex traits from expression data. Bioinformatics 26:896–904

Rossow KD (1998) Porcine reproductive and respiratory syndrome. Vet Pathol 35:1–20

Sahadevan S, Gunawan A, Tholen E, Grosse-Brinkhaus C, Tesfaye D, Schellander K, Hofmann-Apitius M, Cinar MU, Uddin MJ (2014) Pathway based analysis of genes and interactions influencing porcine testis samples from boars with divergent androstenone content in back fat. PLoS One 9, e91077

Sang Y, Brichalli W, Rowland RR, Blecha F (2014) Genome-wide analysis of antiviral signature genes in porcine macrophages at different activation statuses. PLoS One 9, e87613

Schook LB, Beever JE, Rogers J, Humphray S, Archibald A, Chardon P, Milan D, Rohrer G, Eversole K (2005) Swine Genome Sequencing Consortium (SGSC): a strategic roadmap for sequencing the pig genome. Comp Funct Genomics 6:251–255

Schroyen M, Tuggle CK (2015) Current transcriptomics in pig immunity research. Mamm Genome 26:1–20

Schroyen M, Steibel J, Koltes J, Choi I, Raney N, Eisley C, Fritz-Waters E, Reecy J, Dekkers J, Rowland R, Lunney J, Ernst C, Tuggle C (2015) Whole blood microarray analysis of pigs showing extreme phenotypes after a porcine reproductive and respiratory syndrome virus infection. BMC Genomics 16:516

Schroyen M, Eisley C, Koltes JE, Fritz-Waters E, Choi I, Plastow GS, Guan L, Stothard P, Bao H, Kommadath A, Reecy JM, Lunney JK, Rowland RR, Dekkers JC, Tuggle CK (2016) Bioinformatic analyses in early host response to Porcine Reproductive and Respiratory Syndrome virus (PRRSV) reveals pathway differences between pigs with alternate genotypes for a major host response QTL. BMC Genomics 17:196

Serão NV, Matika O, Kemp RA, Harding JC, Bishop SC, Plastow GS, Dekkers JC (2014) Genetic analysis of reproductive traits and antibody response in a PRRS outbreak herd. J Anim Sci 92:2905–2921

Shannon P, Markiel A, Ozier O, Baliga NS, Wang JT, Ramage D, Amin N, Schwikowski B, Ideker T (2003) Cytoscape: a software environment for integrated models of biomolecular interaction networks. Genome Res 13:2498–2504

Sharma A, Lee JS, Dang CG, Sudrajad P, Kim HC, Yeon SH, Kang HS, Lee SH (2015) Stories and challenges of genome wide association studies in livestock – a review. Asian-Australas J Anim Sci 28:1371–1379

Shen-Orr SS, Tibshirani R, Khatri P, Bodian DL, Staedtler F, Perry NM, Hastie T, Sarwal MM, Davis MM, Butte AJ (2010) Cell type-specific gene expression differences in complex tissues. Nat Methods 7:287–289

Shoemaker JE, Lopes TJ, Ghosh S, Matsuoka Y, Kawaoka Y, Kitano H (2012) CTen: a web-based platform for identifying enriched cell types from heterogeneous microarray data. BMC Genomics 13:460

Souza C, Choi I, Araujo K, Abrams S, Kerrigan M, Rowland RR, Lunney J (2013) Comparative serum immune responses of pigs after a challenge with porcine reproductive and respiratory syndrome virus (PRRSV). In: Proceedings of the 10th IVIS International Veterinary Immunology Symposium, P05.14 Milan

Stear MJ, Bishop SC, Mallard BA, Raadsma H (2001) The sustainability, feasibility and desirability of breeding livestock for disease resistance. Res Vet Sci 71:1–7

Subramanian A, Tamayo P, Mootha VK, Mukherjee S, Ebert BL, Gillette MA, Paulovich A, Pomeroy SL, Golub TR, Lander ES, Mesirov JP (2005) Gene set enrichment analysis: a knowledge-based approach for interpreting genome-wide expression profiles. Proc Natl Acad Sci U S A 102:15545–15550

Suchodolski JS, Markel ME, Garcia-Mazcorro JF, Unterer S, Heilmann RM, Dowd SE, Kachroo P, Ivanov I, Minamoto Y, Dillman EM, Steiner JM, Cook AK, Toresson L (2012) The fecal microbiome in dogs with acute diarrhea and idiopathic inflammatory bowel disease. PLoS One 7, e51907

Supek F, Bosnjak M, Skunca N, Smuc T (2011) REVIGO summarizes and visualizes long lists of gene ontology terms. PLoS One 6, e21800

The FAANG Consortium, Andersson L, Archibald AL, Bottema CD, Brauning R, Burgess SC, Burt DW, Casas E, Cheng HH, Clarke L, Couldrey C, Dalrymple BP, Elsik CG, Foissac S, Giuffra E, Groenen MA, Hayes BJ, Huang LS, Khatib H, Kijas JW, Kim H, Lunney JK, McCarthy FM, McEwan JC, Moore S, Nanduri B, Notredame C, Palti Y, Plastow GS, Reecy JM, Rohrer GA, Sarropoulou E, Schmidt CJ, Silverstein J, Tellam RL, Tixier-Boichard M, Tosser-Klopp G, Tuggle CK, Vilkki J, White SN, Zhao S, Zhou H (2015) Coordinated international action to accelerate genome-to-phenome with FAANG, the Functional Annotation of Animal Genomes project. Genome Biol 16:57

Tuggle CK, Towfic F, Honavar VG (2011) Introduction to Systems Biology for Animals Scientists. In: Te Pas MFW WH, Bannink A (eds) Systems biology and livestock science. Wiley-Blackwell, Oxford, pp 1–30

Uddin MJ, Cinar MU, Grosse-Brinkhaus C, Tesfaye D, Tholen E, Juengst H, Looft C, Wimmers K, Phatsara C, Schellander K (2011) Mapping quantitative trait loci for innate immune response in the pig. Int J Immunogenet 38:121–131

Uthe JJ, Wang Y, Qu L, Nettleton D, Tuggle CK, Bearson SM (2009) Correlating blood immune parameters and a CCT7 genetic variant with the shedding of Salmonella enterica serovar Typhimurium in swine. Vet Microbiol 135:384–388

Uthe JJ, Bearson SM, Qu L, Dekkers JC, Nettleton D, Rodriguez Torres Y, O'Connor AM, McKean JD, Tuggle CK (2011) Integrating comparative expression profiling data and association of SNPs with Salmonella shedding for improved food safety and porcine disease resistance. Anim Genet 42:521–534

Van Reeth K, Labarque G, Nauwynck H, Pensaert M (1999) Differential production of proinflammatory cytokines in the pig lung during different respiratory virus infections: correlations with pathogenicity. Res Vet Sci 67:47–52

Videnska P, Sisak F, Havlickova H, Faldynova M, Rychlik I (2013) Influence of Salmonella enterica serovar Enteritidis infection on the composition of chicken cecal microbiota. BMC Vet Res 9:140

Waide EH, Tuggle CK, Serão N, Schroyen M, Hess A, Rowland RRR, Lunney JK, Plastow G, Dekkers JCM (submitted) Genome wide association analysis of piglet response to infection with two porcine reproductive and respiratory syndrome virus isolates. J Anim Sci 87:1638–1647

West CE, Ryden P, Lundin D, Engstrand L, Tulic MK, Prescott SL (2015) Gut microbiome and innate immune response patterns in IgE-associated eczema. Clin Exp Allergy 45:1419–1429

Wimmers K, Murani E, Schellander K, Ponsuksili S (2009) QTL for traits related to humoral immune response estimated from data of a porcine F2 resource population. Int J Immunogenet 36:141–151

Wysocki M, Chen H, Steibel JP, Kuhar D, Petry D, Bates J, Johnson R, Ernst CW, Lunney JK (2012) Identifying putative candidate genes and pathways involved in immune responses to porcine reproductive and respiratory syndrome virus (PRRSV) infection. Anim Genet 43:328–332

Xia J, Benner MJ, Hancock RE (2014) NetworkAnalyst--integrative approaches for protein-protein interaction network analysis and visual exploration. Nucleic Acids Res 42:W167–W174

Xiao S, Wang Q, Jia J, Cong P, Mo D, Yu X, Qin L, Li A, Niu Y, Zhu K, Wang X, Liu X, Chen Y (2010) Proteome changes of lungs artificially infected with H-PRRSV and N-PRRSV by two-dimensional fluorescence difference gel electrophoresis. Virol J 7:107

Xing J, Xing F, Zhang C, Zhang Y, Wang N, Li Y, Yang L, Jiang C, Wen C, Jiang Y (2014) Genome-wide gene expression profiles in lung tissues of pig breeds differing in resistance to porcine reproductive and respiratory syndrome virus. PLoS One 9, e86101

Zhang H, Guo X, Ge X, Chen Y, Sun Q, Yang H (2009) Changes in the cellular proteins of pulmonary alveolar macrophage infected with porcine reproductive and respiratory syndrome virus by proteomics analysis. J Proteome Res 8:3091–3097

Zhou P, Zhai S, Zhou X, Lin P, Jiang T, Hu X, Jiang Y, Wu B, Zhang Q, Xu X, Li JP, Liu B (2011) Molecular characterization of transcriptome-wide interactions between highly pathogenic porcine reproductive and respiratory syndrome virus and porcine alveolar macrophages in vivo. Int J Biol Sci 7:947–959

Zhu L, Yang S, Tong W, Zhu J, Yu H, Zhou Y, Morrison RB, Tong G (2013) Control of the PI3K/Akt pathway by porcine reproductive and respiratory syndrome virus. Arch Virol 158:1227–1234

Zimmerman J (2003) Historical Overview of PRRS virus. In: Zimmerman J, Yoon KJ, Neumann EJ (eds) 2003 PRRS compendium producer edition: a reference for pork producers. National Pork Board, Des Moines, pp 2–7

Computational Methods for Quality Check, Preprocessing and Normalization of RNA-Seq Data for Systems Biology and Analysis

Gianluca Mazzoni and Haja N. Kadarmideen

Abstract

The use of RNA sequencing (RNA-Seq) technologies is increasing mainly due to the development of new next-generation sequencing machines that have reduced the costs and the time needed for data generation.

Nevertheless, microarrays are still the more common choice and one of the reasons is the complexity of the RNA-Seq data analysis. Furthermore, numerous biases can arise from RNA-Seq technology, and these biases have to be identified and removed properly in order to obtain accurate results.

Nowadays, many tools have been developed which allow to perform each step without high-level programming skills. However, each step of the pipeline needs to be performed carefully and requires a good knowledge of both the technology and the algorithms.

In this comprehensive review, we describe the fundamental steps of the pipeline for RNA-Seq analysis to identify differentially expressed genes: raw data quality control, trimming and filtering procedures, alignment, postmapping quality control, counting, normalization and differential expression test.

For each step, we present the most common tools and we give a complete description of their main characteristics and advantages focusing on the statistics that they perform and the assumptions that they make about the data.

The choice of the right tool can have a big impact on the final results. Until now, no gold standard has been established for this type of analysis.

G. Mazzoni (✉) • H.N. Kadarmideen
Department of Large Animal Sciences, University of Copenhagen,
Groennegaardsvej 7, Frederiksberg C 1870, Denmark
e-mail: gianluca.mazzoni@sund.ku.dk

H.N. Kadarmideen (ed.), *Systems Biology in Animal Production and Health, Vol. 2*,
DOI 10.1007/978-3-319-43332-5_3

In conclusion, this review can be useful for both educational purposes as well as for less experienced practitioners of animal genomic research. In the absence of a commonly accepted standard procedure, the general overview presented in this review can help to make the best choices during the implementation of an RNA-Seq pipeline.

1 Introduction

Next-generation sequencing (NGS) technologies allow the generation of huge quantities of biological data. The development of new NGS machines has led to a reduction in costs and the time needed for data generation. In transcriptomics, the use of RNA sequencing technologies is ever increasing. RNA-Seq has considerably more benefits than microarray technology: it does not rely on previous knowledge and annotation, it has a wide range of sensitivity in detecting transcripts and it allows to quantify expression of different isoforms, study specific allele expression and identify new transcripts (Zhao et al. 2014).

The advantages on RNA sequencing compared to microarray technologies are even more valuable in systems genetics and system biologies studies.

RNA sequencing data facilitates delving into the analysis and extracting information about biological pathways and gene function.

Nevertheless, microarrays are still the more common choice for gene expression profiling and for differentially expressed genes analysis. The reasons are many.

The cost is still significantly higher for RNA-Seq than microarrays. Furthermore, RNA-Seq data brings with it logistic challenges, for example, the high storage capacity needed for the huge quantity of raw data produced as well as the computational power needed to perform some steps of the bioinformatics pipeline (Zhao et al. 2014).

Furthermore, RNA-Seq data is more complex, and a good knowledge of the technology and its related aspects are necessary in order to produce reliable results.

Different biases and artifacts that arise from these technologies and specific statistics have to be applied to obtain consistent and reliable results.

Nowadays, there are many tools available to perform all the different steps of the bioinformatics pipeline of RNA-Seq data (Garber et al. 2011). Some of them have a graphical interface which allows researchers with a basic computational background to perform all the steps to the final results. However, a good knowledge of the algorithm and a computational background is still necessary to obtain accurate results and make the correct choices in term of tools and statistical tests. Tools differ in the statistics that they perform and in the assumptions that they make about the data. Therefore, they can be more or less efficient with regard to specific characteristics of the dataset as well as the experimental design.

The basic steps of the bioinformatics pipeline for RNA-Seq data are: raw data quality control followed by trimming and filtering procedures, alignment, postmapping quality control, counting and normalization statistic test for differential expression (Mutz et al. 2013) (Fig. 1).

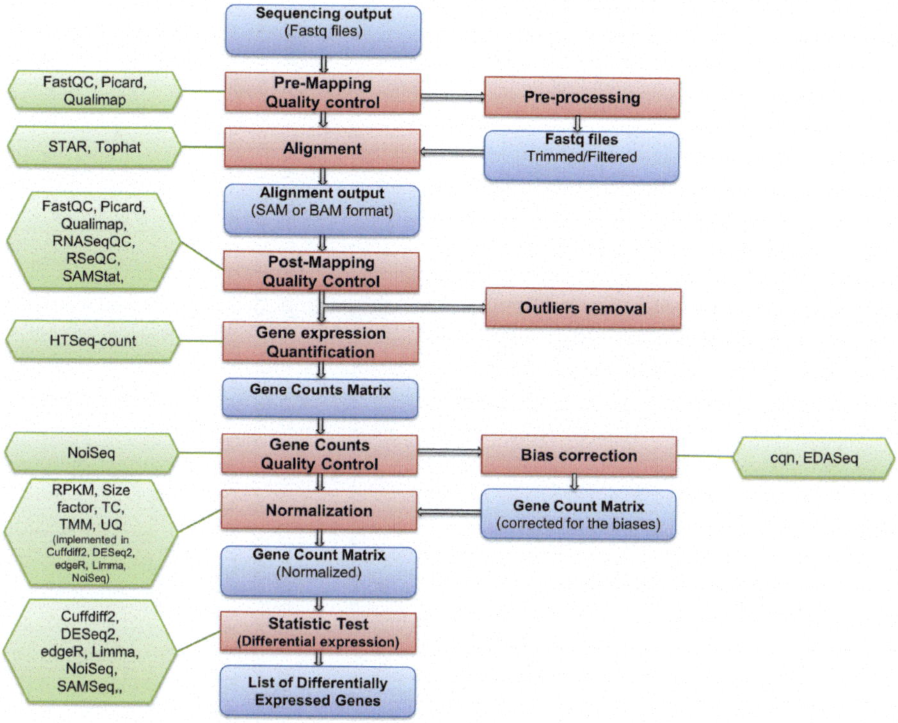

Fig. 1 This picture represents the basic RNA-Seq data analysis pipeline. The *red boxes* are the main steps. The *blue boxes* describe the type of file that is given as input or produced as output at each step. The *green boxes* contain the list of the tools described in the text and they are connected to the step that they perform

2 Raw Data Quality Control

Raw data from RNA-Seq technology is a text file with a FASTQ format. The biological sequences of the reads as well as the sequencing quality values at each nucleotide base are stored in this file. Sequencing quality changes along the positions of the reads usually with a machine specific trend (Fig. 2a).

This bias together with contaminations of unwanted reads and PCR artifacts, GC content and presence of adapters represent technical biases.

Quality control of the raw data is a very important step that facilitates the detection of biases generated during the sequencing procedure that, if not correctly removed, can generate problems like incorrect mapping during the alignment and affect the final results.

The more common tools used in this step are FastQC (Andrews 2010), Qualimap (García-Alcalde et al. 2012) and Picard Tools (Wysoker et al. 2012). These tools are easy to use and the first two also have graphical interfaces for users with no computational skills. The statistics that are usually considered at this step are: total number

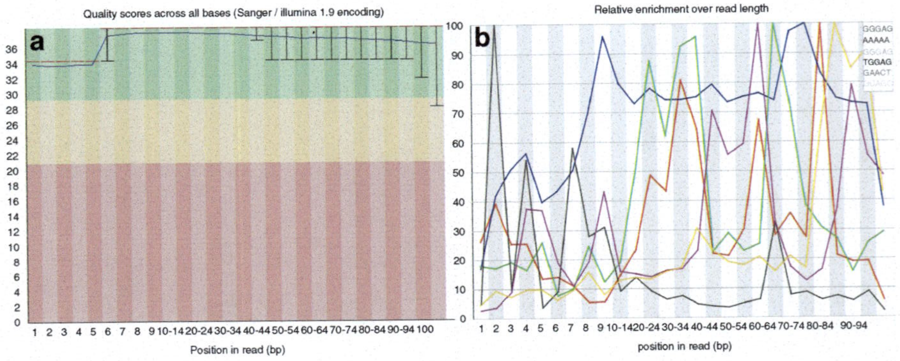

Fig. 2 (**a**) Per base sequence quality plot obtained with FastQC. The RNA sample obtained from bovine cumulus cells has been sequenced with Illumina technology. The reads with low average quality have been filtered out, but it is still possible to see the typical trend where the quality tends to decrease moving along the read length. (**b**) Kmer content computed with FastQC. The *plots* represent the top overrepresented kmers in the sample, across the read positions

of reads, per base sequence quality, per sequence quality score, per base sequence content, per sequence GC content, per base N content, sequence duplication levels, overrepresented sequences and kmer content. This type of quality control is the same as that applied to DNA sequencing data. It is not RNA-specific and it can only provide information about the quality of read data related to NGS technologies.

Bases with low sequencing quality have a higher probability to be wrong. Regions where the quality is too low could have many mistakes that occurred during the sequencing and should be trimmed or filtered out. On the other hand, (Williams et al. 2016) recently found that a too aggressive trimming of RNA-Seq data before gene expression quantification can have great impact on the final estimation leading to unpredictable changes, mainly caused by the generation of very short reads.

Tools like Picard, FastQC or Qualimap compute the summary statistics at each position considering a representative subset of the reads. They generate a boxplot for each position of the read to represent the distribution of the quality per position.

Once identified, this type of issue can be removed by trimming specific regions or entire reads, considering different criteria chosen on the basis of the quality trend of the library.

GC content distribution and overrepresented sequence statistics point out the presence of contaminations or PCR artifacts, or problems during the library preparation.

If the library preparation is carried out correctly, it is expected to have a specific distribution of GC across the set of reads. If the distribution is different from the expected one, it is because there is an overrepresentation probably due to contaminations.

With regard to the level of contamination, if most of the library is represented by contaminations, the sample should be removed, but first it would be better to test whether it is an outlier by using clustering techniques or exploratory analysis such as principal component analysis (PCA). Otherwise, if the contamination represents only a small portion of the library and the sample does not turn out to be an outlier, the contamination can be identified and removed before proceeding with the analysis.

The kmer content is another way to identify biases due to the sequencing or the library preparation technology. The graph represents the overrepresentation of specific sub-sequences along the length of the reads. Library protocols based on random priming have a specific imbalance at the start of the library (Fig. 2b).

Overrepresented reads in the library can be due to strongly expressed transcripts, contaminations, PCR artifacts, adapter content or DNA sequences used during the lab work. Furthermore, they can represent rRNA transcripts that have not been correctly depleted during the RNA purification step. To identify the origin of the overrepresented reads, the sequences can be aligned against RNA sequences in publicly available databases using BLAST or compared against UniVec, an annotated database for vector sequences provided by NCBI (Cochrane and Galperin 2010).

3 Alignment

During this step, the read sequences of the cDNA fragments originating from the random fragmentations and retrotranscription of RNA transcripts are aligned to reference genomes (Wang et al. 2009).

In this way, it is possible to identify the gene or the genomic locus that gave origin to the transcript from which each fragment derived.

The choice of the aligner has to be made considering the library and sequencing protocol as well as the objective of the analysis.

Tophat (Kim and Salzberg 2011) and STAR (Dobin et al. 2013) are two aligners specific for RNA-Seq data (able to identify splicing sites), which have shown the best performances.

Tophat and STAR have been tested together with other aligners using different datasets and they showed similar accuracy (Engström et al. 2013), but the latter has the advantage of being much faster and in the case of large datasets, it can be the best solution.

In the case of de novo mapping, the reads are used to generate contigs and reconstruct the set of isoforms for a specific gene present in a sample directly from the sequenced reads. The process can be performed by using a reference or based only on the reads (Garber et al. 2011).

The set of contigs obtained can then be used as a reference to count the reads that map on them and quantify their expression in the sample. A well-known tool for de novo mapping is Trinity.

Trinity is composed of three independent software modules: Inchworm, Chrysalis and Butterfly. As a final output, the tool gives a full-length transcript with the corresponding alternatively spliced isoforms (Grabherr et al. 2011).

4 Postmapping Quality Control

Postmapping quality control is a fundamental step that allows to identify issues that have occurred during the sequencing or sample extraction or library preparation that can be identified only after alignment.

Nowadays, there are many freely available tools that are able to perform postmapping quality control.

These tools do not have a direct impact on the final results; however, it is fundamental to check the samples before proceeding with the other steps of the pipeline (Williams et al. 2014). Some of the tools are very user-friendly and furthermore, they generate easily interpretable outputs compared to others that need more computational skills.

The most widely used tools are FastQC, Picard Tools, Qualimap, RNA-SeqQC (DeLuca et al. 2012), RSeQC (Wang et al. 2012) and SAMStat (Lassmann et al. 2011).

During the postmapping quality control, two main types of statistics can be performed: general statistics similar to the one applied to raw data and RNA-Seq specific statistics. The first type focuses on NGS-related problems (number of reads

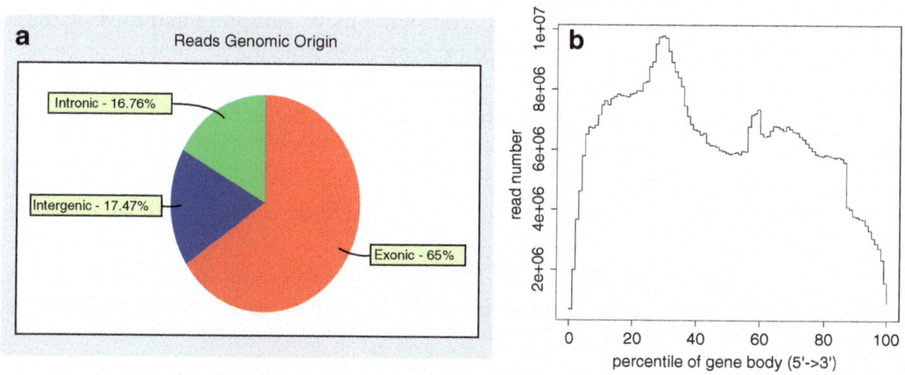

Fig. 3 (**a**) Pie chart obtained with Qualimap showing the percentages of reads mapped to exonic, intronic and intergenic regions of RNA-Seq data from bovine samples. The computation is based on a General Feature Format file where all the information about genomic features of the species of interest were annotated; in this case, we used *Bos taurus* UMD v.3.1.83. (**b**) Gene body coverage computed with RSeQC. The plot represents the coverage along the length of all the transcripts annotated in the bovine genome, normalized from 1 to 100. The reduction present at the 3′ of the transcript indicates a low level of degradation present in the sample

mapped, nucleotide composition, GC percentage, kmer bias) with the only difference being that the statistics are based only on uniquely mapped reads.

FastQC and Picard Tools can also be used at this point of the analysis together with other tools like SAMStat.

SAMStat performs a deeper analysis to detect possible biases related to the mapping quality.

This tool generates a plot where the properties of unmapped, poorly mapped and accurately mapped reads are compared in order to see if some differences are related to the quality of the alignment.

RNA-Seq postmapping statistics focus on genome coverage, intron/exon coverage, intron/exon junction analysis, and in the case of paired end protocols the insert size distribution (Fig. 3a).

Considering that our reads are generated mainly from processed transcripts, especially in the case of mRNA-enriched libraries, we expect that most of them will map to previously annotated exonic regions related to intronic and even less intergenic regions.

These types of statistics are organism-specific because they are strictly dependent on the level of annotation of the genome and obviously on the library protocol used.

Unexpected percentages of reads from intronic and intergenic regions point out problems during library preparation or contamination.

Another important analysis is the intron/exon junction percentages (known, partially known, novel junction). If the sequencing is deep enough and is a good representation of the sample, the spliced junctions should be rediscovered in an RNA-Seq experiment. Spliced junction saturation analysis is also implemented in RSeQC.

The introns/exons junction saturation is computed by re-sampling and thus increasing the total number of reads; thereby computing each time the percentage of known junctions identified.

This information is dependent on the annotation of the genome, but it is important to understand whether the information contained in the data is enough to perform differential splicing.

In the case of paired end protocols, the insert size distribution can be useful to check if the alignment ran correctly.

This statistic has to be specific for RNA, because to compute the correct insert size distribution, the presence of introns when the paired reads are mapped back to the genome have to be considered.

If the sequenced fragment has originated from two exons and the splicing site is in the middle between the forward and the reverse reads, the real insert size can be obtained by subtracting the length of the intron from the distance between the reads' mapping site in the genome.

Insert size statistics are implemented in Picard Tool, Qualimap, RNA-SeqQC and RSeQC. While the first three extract it directly from the SAM file, RSeQC performs a more complex computation, taking the possible presence of introns between two paired reads into consideration.

RSeQC and Qualimap are able to compute an interesting postmapping quality control called gene body coverage. This test is useful, especially in cases where samples have problems in the quality and integrity of the RNA.

The tools give as output a graph representing the level coverage across the length of the transcripts present in the genomes, normalized from 1 to 100 (Fig. 3b).

Qualimap, together with Picard Tools, provides a module specific for RNA sequencing and together with RSeQC and RNA-SeqQC represent the most complete tools for postalignment quality control in RNA-Seq data.

RNA-SeqQCs can also perform a multisample comparison providing information such as correlations and GC content comparisons among samples.

Some tools are less intuitive, while other packages like Qualimap have a well developed graphical interface and provide a complete, well-organized graphical output particularly useful for researchers with weak computational skills.

The ideal way to get a complete impression of the data is to combine the results from different tools, exploiting the advantages of each of them.

This concept is implemented in a recently developed tool called Quality Control for RNA-Seq (QuaCRS) (Kroll et al. 2014). The tool runs FastQC, RNA-SeqQC and SeQC and merges results in an easily interpretable and accessible way.

5 Counting

In this step, reads that map under a biological feature of interest are counted in order to quantify its expression in a sample. Various tools perform this step. The differences are few among these types of tools and they are related mainly in the different ways of considering reads that overlap more than one feature. The estimation of the expression can be made at different levels for different biological features (gene level, transcript level, exonic level), or it can be applied to all the transcripts identified during de novo mapping.

For example, HTSeq (Anders et al. 2014) and Cufflinks (Trapnell et al. 2013) are commonly used tools to perform this step.

6 Normalization

Even if RNA sequencing was initially considered completely immune of biases, normalization is still a fundamental step (Wang et al. 2009).

It facilitates the removal of biases and it is necessary in order to obtain accurate results during the comparison both within and between samples.

Normalization is tricky and complex in RNA-Seq data, as there are different bias types to take into consideration. In RNA-Seq experiments, biases can be of two types: within-sample bias that is due mainly to gene length bias and GC content bias, and between-sample biases due to the sequencing depth (Dillies et al. 2013).

The gene length bias originates because longer transcripts likely generate a higher number of fragments and consequently a higher number of reads. Thus, it is likely to have a higher level of expression rather than shorter transcripts due to this technical problem and not due to a real activation or inactivation of the transcription (Zheng et al. 2011; Oshlack and Wakefield 2009).

Similar problems occur in fragments with different GC contents (Risso et al. 2011).

GC-rich and GC-poor fragments result in being underrepresented in RNA sequencing, which leads to biases at the gene expression level (Benjamini and Speed 2012).

To make things even more complicated, it has been seen that GC content bias is not consistent between samples. It is lane-dependent and probably introduced during the library preparation step (Risso et al. 2011).

Until now, it has not been determined which method performs better in normalizing for GC content bias.

One of the methods used to account for length bias is the RPKM unit (reads per kilobase of exon per million fragments), which divides the discrete counts of the reads by the total number of reads sequenced and by the length of the transcript and then computes the proportion to one million total reads (Mortazavi et al. 2008). In this way, the expression value of a gene is independent on the length of its transcripts.

Various tools are able to correct for this type of bias, like EDASeq (Risso et al. 2011) and cqn (Hansen et al. 2012) where the GC bias or length bias are included as covariates.

The correction for the biases is dependent on the objective of the study.

If the objective is to rank genes within a sample, for example, to identify which genes are more active in a specific cell type, the biases that must be checked are gene length and GC content.

On the other hand, if the experiment is designed to compare gene expressions between samples to identify differentially expressed genes, the most influential bias to consider is the difference in the library size.

The library size, computed as the total number of reads in a sample, can lead to false positives or false negatives during the analysis, as more reads will be assigned to each gene if a sample is sequenced to a greater depth.

However, it is also very important to consider that gene length and GC content also have an effect in between-sample comparisons; in fact, genes with higher counts are more likely to be defined as differentially expressed than genes with lower counts.

Cqn and EDASeq have been developed in such a way that they correct first for within-sample effect of the GC content and then they correct for between-sample bias.

It has been seen that normalization for library size with simple scaling is not enough. Together with sequencing depth and gene length, the composition of the RNA population has to be considered.

If the majority of genes are highly expressed in one condition compared to the other, the results of the analysis will be skewed (Robinson and Oshlack 2010).

More sophisticated normalization methods have been developed to correct for differences in library size (Oshlack et al. 2010).

Normalization methods have been tested with different datasets (Dillies et al. 2013).

The method implemented in the DESeq2 package (Anders and Huber 2010), together with trimmed mean of M values (TMM) (Robinson and Oshlack 2010) showed good precision and sensibility in false positive rates and power of detection. The first methods use scaling factor for each sample, computed as the median of the ratio between genes and their respective geometric mean computed across samples, while TMM removes the genes that are most expressed and with the highest log ratios and using the remaining genes, a scaling factor is computed as the weighted mean of log ratios between the sample and a reference.

Other methods are also used with good performances, such as upper quartile (Bullard et al. 2010), where gene counts are divided by the upper quartile of the gene counts and median where gene counts are divided by the median of the gene counts.

Even if RPKM, as explained earlier, takes into account the gene length, this method together with total count (TC), in which the counts of the genes are divided by the total number of reads in the sample, is indicated to be ineffective.

The performance of a normalization method is strictly dependent on the dataset. In some cases, no differences have been found in the final results between various methods (Seyednasrollah et al. 2015).

In general, there is no agreement on which is the best method and it is very important to check if the normalization applied worked fine on a dataset. This can be achieved by comparing the median and the distribution of gene expression across genes. In this way, it is possible to identify batch effect on the samples. We expect that after normalization, if the procedure is performed correctly, the distributions should have similar medians and distributions across samples. A similar test is

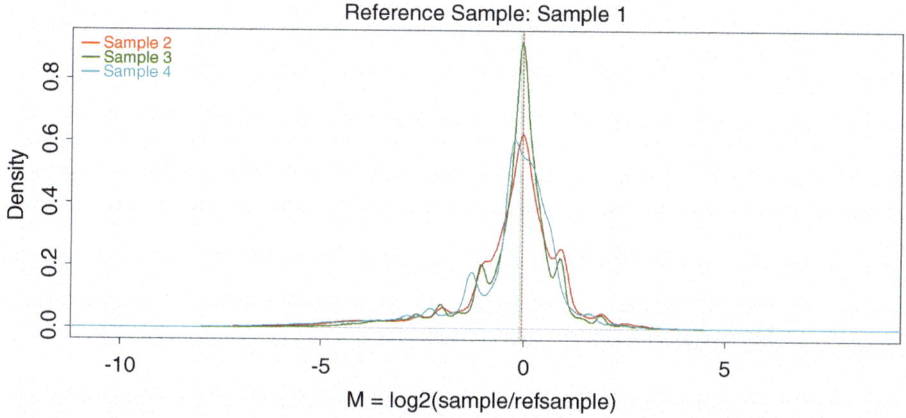

Fig. 4 NOISeq batch effect exploration graph. A sample is used as a reference and NOISeq compares the distributions and the medians among all the samples. The RNA samples analyzed are obtained from bovine cumulus cells, sequenced with Illumina technology using the same library preparation. The samples need to be normalized before proceeding with the next step of the analysis. This issue is mainly due to differences in library sizes

provided by NOISeq (Tarazona et al. 2012). This R package compares read distributions among samples using a sample as a reference, check for presence of GC content bias and length bias (Fig. 4).

Once the data are correctly normalized and transformed, various exploratory analyses can be performed and systems genetic approaches can be applied.

Normalization has less influence in the case of co-expression analysis because we focus on the correlations between expression levels of pairs of genes across all the samples.

In any case, tools for co-expression analysis suggest normalizing the data and applying logarithm-based transformations. For example, WGCNA (Langfelder and Horvath 2008) suggests using variance stabilizing transformation of RNA-Seq data before proceeding with the analysis.

7 Statistical Analysis

At this point of the pipeline, data appear in a matrix where each entry represents the expression level for a gene in one sample.

The normalized matrix can be used as input for the following steps of the analysis: differential expression, co-expression analysis or exploratory analysis like clustering and data visualization.

At this point, the normalized matrix can be treated in the same way as matrices originating from microarray technologies.

One difference has to be taken into consideration: values from RNA-Seq data are discrete measures because they are based on counts of the reads, while microarray data are continuous measures based on intensity values (Fang et al. 2012).

RNA-Seq data are characterized by two properties: the presence of extreme values and heteroscedasticity (relation between variance and mean of gene expressions).

For these reasons, data from RNA-Seq data are usually transformed in a logarithmic way or with other types of transformation like variance stabilizing transformation (Lin et al. 2008). Tools developed in a specific way for RNA sequencing do not need logarithmic transformation because they already take into account the typical distribution of the data counts.

8 Differential expression analysis

DE analysis allows to recognize genes whose expression is related to a trait of interest, such as those genes whose expression changes between conditions with enough statistical power. In this step, a statistical test is applied to each gene to determine whether we have enough statistical power to reject the null hypothesis that the gene is equally expressed in two or more conditions.

Differentially expressed genes provide information about the functions of genes under different conditions. From a systems biology perspective, the analysis of a set of DE genes can be integrated with information from different omics levels, leading to the identification of potential biological pathways involved in a process.

In RNA-Seq, this step is one of the most critical, for which a number of methods have been developed.

Each method is based on different assumptions regarding the distribution of the gene counts and on different statistical models. Some of them can deal with multifactorial analysis, others can be applied in experimental designs with no replicates, while still others allow for isoform detection and quantification (Mazzoni et al. 2015). Above all, the performances are dependent on the structure of the data.

Many tools have been tested with both real and simulated data sets. From these studies, the performances of the tools are strictly dependent on the properties of the dataset and on the experimental design (Zhang et al. 2014; Seyednasrollah et al. 2015).

The choice of the tool is fundamental. Taking into consideration that there is great variability in the maturity (Garber et al. 2011) of available computational tools, it is important that the user is aware of the main differences and makes a choice considering properties of the data like number of samples, replicates and heterogeneity of the dataset (Seyednasrollah et al. 2015).

Tools for differential expression can be classified in non-parametric tools that are not based on the assumption of the distribution of the gene counts, and the parametric tool where gene expression of the genes is assumed to have a specific distribution.

Among the non-parametric methods we find NOISeq and SAMSeq.

Both of them perform very well in terms of control of false positives, but they have opposite characteristics: NOISeq is too conservative with a high number of replicates, while SAMSeq needs more replicates for a good power of detection and its performances are strictly related to the data (Soneson and Delorenzi 2013; Seyednasrollah et al. 2015).

Among parametric methods, the best performing tools are DESeq, edgeR (Robinson et al. 2010) and BaySeq (Hardcastle and Kelly 2010), which appear to be similar in terms of accuracy, control of the number of false positives and sensitivity (Zhang et al. 2014; Kvam et al. 2012).

In datasets with small sample size, the best tools turned out to be Limma and DESeq.

DESeq proved to be the most conservative, while edgeR has a higher power of detection and Limma is the most robust with strong consistency of the results across heterogeneous datasets (Seyednasrollah et al. 2015; Soneson and Delorenzi 2013).

DESeq's successor, DESeq2, has a higher power of detection, but is less precise (Seyednasrollah et al. 2015).

BaySeq, based on Bayesian methodology, showed good performances in different cases but is strongly dependent on the dataset structure (Seyednasrollah et al. 2015; Soneson and Delorenzi 2013).

Finally, one of the most prominent tools, Cuffdiff2, has good performances but poor power of detection at the gene level (Seyednasrollah et al. 2015; Zhang et al. 2014).

However, one of the main advantages of Cuffdiff2 is the possibility to compute expression changes at the gene and transcript levels.

In the case of complex experimental designs, where more than one variable can be correlated to the gene expression levels, the possibility of accounting for those variables in the model is very important.

DESeq, DESeq2, edgeR, Limma and NOISeq allow for performing multifactorial analysis (Love et al. 2014; Robinson et al. 2010; Ritchie et al. 2015; Tarazona et al. 2012). Thanks to these tools, it is very easy to deal with very complex experimental designs, even for less experienced users.

Typically, the user gives as input the linear model that the tool will fit before computing the contrast. The basic model is:

$$y_i = \text{covariate}_1 + \text{covariate}_2 + \text{covariate}_n + \text{trait\_of\_interest}$$

where y_i is the gene normalized gene counts for gene i across all the samples, covariate 1 to n represents potential confounding effects that have to considered during the test and the trait of interest is the covariate, which has to be performed for the differential expression analysis.

The program will fit many models as the number of genes given in input ($i = 1$ to t), where t is the number of genes to be tested.

For DESeq, edgeR and Limma, very extensive explanations of the tools are provided together with the manuals, making their use and the interpretation of the results even easier (Seyednasrollah et al. 2015).

```
                     baseMean log2FoldChange      lfcSE       stat    pvalue      padj
                    <numeric>      <numeric> <numeric>  <numeric> <numeric> <numeric>
ENSBTAG00000000010  956660606     0.02814910 0.2112653  0.1332405 0.8940032 0.9976981
ENSBTAG00000000012  212395995    -0.12010838 0.2113485 -0.5682953 0.5698344 0.9976981
ENSBTAG00000000013  273204564    -0.23214247 0.2059338 -1.1272672 0.2596295 0.9851721
ENSBTAG00000000014 1495445436     0.16548246 0.1891041  0.8750867 0.3815267 0.9976981
ENSBTAG00000000015   70768749     0.04718367 0.2770936  0.1702806 0.8647894 0.9976981
...                       ...            ...       ...        ...       ...       ...
ENSBTAG00000048293   53897117    -0.07879336 0.2846027 -0.2768539 0.7818923 0.9976981
ENSBTAG00000048296   74821939    -0.32501805 0.2840713 -1.1441424 0.2525646 0.9851721
ENSBTAG00000048306   21861174    -0.23933382 0.2682570 -0.8921811 0.3722959 0.9976981
ENSBTAG00000048308  166023977    -0.20215832 0.2682399 -0.7536474 0.4510610 0.9976981
ENSBTAG00000048314  187840008    -0.09321117 0.2674218 -0.3485549 0.7274235 0.9976981
```

Fig. 5 DESeq2 results from a differential expression analysis performed on bovine RNA-Seq data. *BaseMean*, mean of normalized counts for all samples; *log2FoldChange*, estimate of the gene expression change for the trait analysed (reported in a $\log^2$ scale); *lfcSE*, standard error associated to the estimate; *stat*, Walt test statistic; *p*-value, *p*-values obtained from the Walt test; *padj*, *p*-values adjusted for multiple testing (Benjamini–Hochberg procedure)

9 Interpretation of DE Analysis Results

The output file generated by most of the tools from a differential expression analysis consists of a list of genes or features followed by different parameters obtained from the statistic tests (Fig. 5).

The important parameters that are obtained from a differential expression analysis and that generally are presented in the final results file are the estimated fold change, the associated *p*-value and the *p*-value adjusted for multiple testing. The estimated fold change is the effect size estimate. The effect size estimate represents how much the expression of a gene changes due to the condition for which the contrast has been computed. Usually, this parameter is in a base 2 logarithmic scale. The tools compute also a statistic test that can be, for example, a Walt test, a likelihood ratio test or a Bayes statistic in order to obtain a *p*-value associated to the estimates.

Together with the *p*-value, the related adjusted *p*-value is also usually computed. The adjusted *p*-value is the statistic significance after multiple testing corrections. Usually the multiple testing is based on false discovery rate, but each tool gives the possibility to choose between different methods (Robinson et al. 2010; Anders and Huber 2010; Ritchie et al. 2015; Love et al. 2014).

The adjusted *p*-values give information about the significance of the gene expression change.

In general, to evaluate the differentially expressed genes, two thresholds should be set up; one for the adjusted *p*-value and another one for the fold change. In this way, it is possible to select genes whose change in expression is statistically significant and with a certain magnitude.

Conclusions

In this review, we have summarized all basic steps of the pipeline for RNA-Seq data analysis focusing on the steps that allow to check and get rid of the biases that can arise from RNA-Seq data.

In order to obtain accurate results, it is really important to remove potential sources of biases. The choice of the right tool, as well as the choice on how to identify problems in the data and to get rid of them, can have big impact on the final results.

This choice is not always easy and in order to perform a good analysis, it requires good knowledge about the tools available as well as about the RNA-Seq technology.

While for microarray analysis, the general standard to record and report microarray-based gene expression data has been defined in the MIAME guideline (Brazma et al. 2001), until now, no golden standard has been described for RNA-Seq data analysis.

One of the objectives of the FAANG project (http://www.faang.org/) is to establish a standard procedures for core assays, experimental protocols and also for RNA-Seq analysis pipeline in animal genomic research field.

In the absence of a commonly accepted standard procedure, the general overview presented in this review can help the reader in setting up the analytic pipeline. Furthermore, it can help to make the best choice in term of tools to use, thanks to the wide description of their characteristic and of the comparison of their performances.

In conclusion, this review can be useful for both educational purposes as well as for less experienced practitioners of animal genomic research who are dealing with RNA-Seq data.

Acknowledgments We thank Programme Commission on Health, Food and Welfare of the Danish Council for Strategic Research (Innovationsfonden) for financial support within the GIFT project.

References

Anders S, Huber W (2010) Differential expression analysis for sequence count data. Genome Biol 11(10):R106

Anders S, Pyl PT, Huber W (2014) HTSeq–A Python framework to work with high-throughput sequencing data. Bioinformatics btu638, 31(2):166–9

Andrews S (2010) FastQC: a quality control tool for high throughput sequence data., Reference Source

Benjamini Y, Speed TP (2012) Summarizing and correcting the GC content bias in high-throughput sequencing. Nucleic Acids Res gks001, 40(10):e72

Brazma A, Hingamp P, Quackenbush J, Sherlock G, Spellman P, Stoeckert C, Aach J, Ansorge W, Ball CA, Causton HC (2001) Minimum information about a microarray experiment (MIAME)—toward standards for microarray data. Nat Genet 29(4):365–371

Bullard JH, Purdom E, Hansen KD, Dudoit S (2010) Evaluation of statistical methods for normalization and differential expression in mRNA-Seq experiments. BMC Bioinform 11(1):94

Cochrane GR, Galperin MY (2010) The 2010 nucleic acids research database issue and online database collection: a community of data resources. Nucleic Acids Res 38(suppl 1):D1–D4

DeLuca DS, Levin JZ, Sivachenko A, Fennell T, Nazaire M-D, Williams C, Reich M, Winckler W, Getz G (2012) RNA-SeQC: RNA-seq metrics for quality control and process optimization. Bioinformatics 28(11):1530–1532

Dillies M-A, Rau A, Aubert J, Hennequet-Antier C, Jeanmougin M, Servant N, Keime C, Marot G, Castel D, Estelle J (2013) A comprehensive evaluation of normalization methods for Illumina high-throughput RNA sequencing data analysis. Brief Bioinform 14(6):671–683

Dobin A, Davis CA, Schlesinger F, Drenkow J, Zaleski C, Jha S, Batut P, Chaisson M, Gingeras TR (2013) STAR: ultrafast universal RNA-seq aligner. Bioinformatics 29(1):15–21

Engström PG, Steijger T, Sipos B, Grant GR, Kahles A, Rätsch G, Goldman N, Hubbard TJ, Harrow J, Guigó R (2013) Systematic evaluation of spliced alignment programs for RNA-seq data. Nat Methods 10(12):1185–1191

FAANG (Functional Annotation of Animal Genomes). http://www.faang.org/

Fang Z, Martin J, Wang Z (2012) Statistical methods for identifying differentially expressed genes in RNA-Seq experiments. Cell Biosci 2(1):26

Garber M, Grabherr MG, Guttman M, Trapnell C (2011) Computational methods for transcriptome annotation and quantification using RNA-seq. Nat Methods 8(6):469–477

García-Alcalde F, Okonechnikov K, Carbonell J, Cruz LM, Götz S, Tarazona S, Dopazo J, Meyer TF, Conesa A (2012) Qualimap: evaluating next-generation sequencing alignment data. Bioinformatics 28(20):2678–2679

Grabherr MG, Haas BJ, Yassour M, Levin JZ, Thompson DA, Amit I, Adiconis X, Fan L, Raychowdhury R, Zeng Q (2011) Full-length transcriptome assembly from RNA-Seq data without a reference genome. Nat Biotechnol 29(7):644–652

Hansen KD, Irizarry RA, Zhijin W (2012) Removing technical variability in RNA-seq data using conditional quantile normalization. Biostatistics 13(2):204–216

Hardcastle TJ, Kelly KA (2010) baySeq: empirical Bayesian methods for identifying differential expression in sequence count data. BMC Bioinformatics 11(1):422

Kim D, Salzberg SL (2011) TopHat-Fusion: an algorithm for discovery of novel fusion transcripts. Genome Biol 12(8):R72

Kroll KW, Mokaram NE, Pelletier AR, Frankhouser DE, Westphal MS, Stump PA, Stump CL, Bundschuh R, Blachly JS, Yan P (2014) Quality control for RNA-seq (QuaCRS): an integrated quality control pipeline. Cancer Inform 13(Suppl 3):7

Kvam VM, Liu P, Si Y (2012) A comparison of statistical methods for detecting differentially expressed genes from RNA-seq data. Am J Bot 99(2):248–256

Langfelder P, Horvath S (2008) WGCNA: an R package for weighted correlation network analysis. BMC Bioinform 9(1):559

Lassmann T, Hayashizaki Y, Daub CO (2011) SAMStat: monitoring biases in next generation sequencing data. Bioinformatics 27(1):130–131

Lin SM, Du P, Huber W, Kibbe WA (2008) Model-based variance-stabilizing transformation for Illumina microarray data. Nucleic Acids Res 36(2):e11–e11

Love MI, Huber W, Anders S (2014) Moderated estimation of fold change and dispersion for RNA-seq data with DESeq2. Genome Biol 15(12):1–21

Mazzoni G, Kogelman L, Suravajhala P, Kadarmideen H (2015) Systems genetics of complex diseases using RNA-sequencing methods. Int J Biosci Biochem Bioinform 5(4):264

Mortazavi A, Williams BA, McCue K, Schaeffer L, Wold B (2008) Mapping and quantifying mammalian transcriptomes by RNA-Seq. Nat Methods 5(7):621–628

Mutz K-O, Heilkenbrinker A, Lönne M, Walter J-G, Stahl F (2013) Transcriptome analysis using next-generation sequencing. Curr Opin Biotechnol 24(1):22–30

Oshlack A, Wakefield MJ (2009) Transcript length bias in RNA-seq data confounds systems biology. Biol Direct 4(1):14

Oshlack A, Robinson MD, Young MD (2010) From RNA-seq reads to differential expression results. Genome Biol 11(12):220

Risso D, Schwartz K, Sherlock G, Dudoit S (2011) GC-content normalization for RNA-Seq data. BMC Bioinform 12(1):480

Ritchie ME, Phipson B, Wu D, Hu Y, Law CW, Shi W, Smyth GK (2015) limma powers differential expression analyses for RNA-sequencing and microarray studies. Nucleic Acids Res gkv007, 43(7):e47

Robinson MD, Oshlack A (2010) A scaling normalization method for differential expression analysis of RNA-seq data. Genome Biol 11(3):R25

Robinson MD, McCarthy DJ, Smyth GK (2010) edgeR: a Bioconductor package for differential expression analysis of digital gene expression data. Bioinformatics 26(1):139–140

Seyednasrollah F, Laiho A, Elo LL (2015) Comparison of software packages for detecting differential expression in RNA-seq studies. Brief Bioinform 16(1):59–70

Soneson C, Delorenzi M (2013) A comparison of methods for differential expression analysis of RNA-seq data. BMC Bioinform 14(1):91

Tarazona S, García F, Ferrer A, Dopazo J, Conesa A (2012) NOIseq: a RNA-seq differential expression method robust for sequencing depth biases. EMBnet J 17(B):18–19

Trapnell C, Hendrickson DG, Sauvageau M, Goff L, Rinn JL, Pachter L (2013) Differential analysis of gene regulation at transcript resolution with RNA-seq. Nat Biotechnol 31(1):46–53

Wang Z, Gerstein M, Snyder M (2009) RNA-Seq: a revolutionary tool for transcriptomics. Nat Rev Genet 10(1):57–63

Wang L, Wang S, Li W (2012) RSeQC: quality control of RNA-seq experiments. Bioinformatics 28(16):2184–2185

Williams AG, Thomas S, Wyman SK, Holloway AK (2014) RNA¯seq data: challenges in and recommendations for experimental design and analysis. Curr Protoc Hum Genet 11.13. 11–11.13. 20

Williams CR, Baccarella A, Parrish JZ, Kim CC (2016) Trimming of sequence reads alters RNA-Seq gene expression estimates. BMC Bioinform 17(1):1

Wysoker A, Tibbetts K, Fennell T (2012) Picard. http://picard.sourceforge.net.

Zhang ZH, Jhaveri DJ, Marshall VM, Bauer DC, Edson J, Narayanan RK, Robinson GJ, Lundberg AE, Bartlett PF, Wray NR (2014) A comparative study of techniques for differential expression analysis on RNA-Seq data 9(8):e103207

Zhao S, Fung-Leung W-P, Bittner A, Ngo K, Liu X (2014) Comparison of RNA-Seq and microarray in transcriptome profiling of activated T cells. PLoS One 9(1)

Zheng W, Chung LM, Zhao H (2011) Bias detection and correction in RNA-Sequencing data. Bmc Bioinform 12(1):290

Systems Biology Application in Feed Efficiency in Beef Cattle

Heidge Fukumasu, Miguel Henrique Santana, Pamela Almeida Alexandre, and José Bento Sterman Ferraz

Abstract

Feed efficiency could be defined as the capacity to generate products with a certain amount of food provided; therefore, the performance and feed intake (FI) are the main components that influence this capacity. In beef cattle, this aims to improve the production both by reducing feed costs, which accounts for a large part of total costs, and by increasing muscle and adipose tissue growth. It is common sense that many physiological processes are involved in the regulation of this trait, such as feed intake, digestion, body composition, metabolism, activity, behavior and thermoregulation. Here, we review the importance of feed efficiency for cattle production, discussing its biological bases from a holistic point of view, finalizing with the possible use of systems biology to improve this important phenotype for animal production.

1 Introduction

Systems biology (SB) could be defined as the study of interactions between the components of biological systems and how these interactions influence the function and behavior of those systems. From a simplified point of view, SB is based on the understanding that the whole is greater than the sum of the parts (www.

H. Fukumasu, DVM, PhD (✉) M.H. Santana, MSc, PhD
P.A. Alexandre, MSc, PhD • J.B.S. Ferraz, DVM, PhD
Departamento de Medicina Veterinária, Faculdade de Zootecnia e Engenharia de Alimentos, Universidade de São Paulo, Pirassununga, São Paulo 13635-900, Brazil
e-mail: fukumasu@usp.br

H.N. Kadarmideen (ed.), *Systems Biology in Animal Production and Health, Vol. 2*,
DOI 10.1007/978-3-319-43332-5_4

systembiology.org). The application of SB in animal production is an emerging and interesting area in animal sciences and will surely lead to the discovery of new hypotheses and tools to improve efficiency and sustainability of this important area. Some work can be found on systems biology in domestic animals, mostly related to growth and reproduction traits (Widmann et al. 2013; Canovas et al. 2014; Kadarmideen 2014). In this chapter, we will discuss systems biology approaches in animal production, using feed efficiency in beef cattle as an example.

2 What Is and Why Feed Efficiency in Beef Cattle?

In the last decade, Brazil and Australia have been the two major beef exporters in the world, but recently, India has become another major player in the market. Although some countries make use of feedlot systems, the vast majority of their animals are produced under pasture conditions, at least during the major part of their lives (Ferraz and Felício 2010; Millen and Arrigoni 2013; Meyer and Rodrigues 2014; Lobato et al 2014). These three countries, as well as the USA, are responsible for almost 2/3 of total beef exports. The herds of Brazil, India and Australia are largely comprised of *Bos indicus* cattle. It is important to highlight that the majority of animals are pasture fed, which is the lowest technology level but the highest in the production of greenhouse gases and methane. Beef cattle production in Latin America accounts for 29% of the world's cattle population and beef production. In addition, Latin America is a region of the world that can significantly increase its production in response to beef demand (Montaldo et al 2012). In experiments that control feed intake individually, it is common to see some animals that eat less than 4 kg of dry matter to gain 1 kg of live weight and others that eat more than 20 kg to gain the same weight. Therefore, it is undeniable that increasing the productivity of that subspecies should be one of the directions toward helping to combat world hunger and reduce the environmental impact of beef production while contributing to a decrease in the production of greenhouse gases. This improvement can be achieved by improving environment (nutrition, reproduction, animal welfare and health) and breeding (fertility, carcass traits, performance and feed efficiency).

The most common way to measure productivity is by the ratio of the amount of resources used for the products generated. In livestock, improved efficiency may be defined as the generation of animal products (meat, milk, wool, eggs, etc.) with a lesser amount of resources or by increasing the generation of products with the same amount of resources already used.

Feed efficiency (FE) means to measure the productivity of animals. FE may be defined as the capacity to generate products with a certain amount of food provided; therefore, the performance and feed intake (FI) are the main components that influence this capacity. The relationship between FI (used resource) and performance (production) resulted in several FE traits (productivity). In beef cattle, this relationship aims to improve the production both by reducing feed costs, which accounts for a large part of total costs, and by increasing muscle and adipose tissue growth (beef cattle). At the same time, the interest in FE also includes concerns about environmental impact, considering its role in reducing the relative greenhouse gas emissions and solid waste (Hegarty et al. 2007; Nkrumah et al. 2006).

Many factors affected breeding cattle for feed efficiency because the priority was for recording production phenotypes but not traits such as feed intake. This can be explained by the cost and complexity of measuring individual FI, so evaluating FE in beef cattle has been concentrated mostly in universities and research centers, which have proper facilities and manpower for such evaluations. However, in recent decades, there has been a considerable increase in commercial genetic selection for FE in beef cattle and, most recently, in *B. indicus* cattle. The adoption of FE selection by animal breeding programs will gradually expand around the world. Concurrently, there is an increase in research on this important characteristic, especially from the point of view of the genetic selection effects over generations and understanding the physiology and genetics behind the difference between the more and less efficient animals.

Many efforts have been made to better balance the relationship between performance and FI in beef cattle. The goal is to find the best way to describe FE effectively and to incorporate it in genetic selection indexes. The selection only for FI is not suitable for *B. indicus* cattle, since it has a high negative correlation with adult weight at maturity (Crews 2005), and the effects of increasing adult weight can be very harmful in extensive production systems based on grazing. Some FE traits (ratio) also turn out to be less effective in reducing this negatively correlated effect with increasing body size, such as feed conversion rate (FCR) and the gross feed efficiency (Kennedy et al. 1993; Crews 2005). Another issue of FCR is that it does not have a real mean and variance is not normally distributed (Atchley and Anderson 1978; Gunsett 1984). In addition, it is highly conditioned to weight gain (Aggrey and Rekaya 2013). Besides these two measures, dozens of other ratio traits were proposed for evaluating the FE of the animals, especially the Kleiber ratio (Kleiber 1947), the partial efficiency of growth and the relative growth rate (Fitzhugh and Taylor 1971). From the point of view of genetic breeding, one of the major problems of using ratio traits in selection indexes (indices) is that these direct ratios are from measures already used in these indices (weight gain), so these linear indices have non-normal distribution; therefore, the estimates of linear equations would become biased (Gunsett 1984; Werf 2004).

To overcome these problems with ratio traits, some measures calculated as residual (regression equations) were proposed or returned to be considered for measurement of FE, especially the residual feed intake (Koch et al. 1963) and the recently proposed residual intake and body weight gain (Berry and Crowley 2012). The residual feed intake (RFI) was proposed in the 1960s and eventually gained more recognition because it is considered phenotypically independent of growth and body size and focused on reducing the FI (Arthur et al. 2001a, b). This phenotypic independence is due to the fact that the RFI is calculated as the difference between observed and estimated FI (based on maintenance of body weight and performance). Despite the good acceptance of the RFI, no measure of FE is definitive and there is no consensus on how to use this information in the selection process (Rolfe et al. 2011). In *B. indicus* cattle, the challenges are even greater because of the small number of animals evaluated and limited knowledge about potential consequences of genetic selection in later generations (Barwick et al. 2009; Grion et al. 2014; Santana et al. 2014a).

The evaluation of feed efficiency is generally performed in young growing animals, so the problem presented by the low number of phenotypes obtained adds to the challenge of estimating breeding values for these animals with high accuracy (low or non-existent progenies). One way to deal with the young age and the relatively low number of evaluated animals enabling a selection scheme under these conditions is by using molecular marker information via genomic selection. The inclusion of these markers can increase the accuracy of estimated breeding values of young animals and thus accelerate the process of genetic selection by reducing the interval between generations.

Even though such assessment is usually made in younger animals, the benefits of improvement in FE can be observed in every phase of beef cattle growth. There is evidence that selection for FE extends to all stages of the production. More efficient animals in the calving phase also showed better FE in finishing (Arthur et al. 2001a, b) and dams with better FE in growing had lower FI in the adult stage with no differences in reproductive rates (Arthur et al. 2005; Basarab et al. 2007). Basically, at the stage of calving, the efficient dams can reduce FI and ensure their requirements for maintenance, reproduction and milk production to assure proper growth of their offspring annually. After weaning, at the phase of growing and finishing, the goal is to guarantee the muscle and lipid tissue growth until the animal reaches the ideal characteristics for slaughter. However, to understand the biology of FE, a holistic view is needed because these traits are very complex and genetics alone is insufficient to understand this biology.

3 A Holistic View of Biology of Feed Efficiency

Although the biological basis for individual variation in feed efficiency has not been fully elucidated, it is common sense that many physiological processes are involved in the regulation of this trait, such as feed intake, digestion, body composition, metabolism, activity and thermoregulation (Basarab et al. 2003; Herd and Arthur 2009; Herd et al. 2004). For many years, studies have focused on analyzing various specific aspects of this trait and we now know that animals with divergent phenotypes for feed efficiency also differ in temperament, feeding behavior, response to stress, appetite, fat deposition, oxygen consumption, energy expenditure, heat loss, mitochondrial function and others.

Using evidence from studies in cattle, chicken, pigs and mice, some authors have discussed that increased physical activity, ingestion behavior and stress are associated with lower efficiency, and that they lead to greater metabolic rate and energy consumption (Herd and Arthur 2009; Herd et al. 2004). Recent studies in cattle also corroborate those results by reporting differences in feed efficiency between animals classified as "adequate temperament" and "excitable temperament" (Francisco et al. 2015); flightier animals present higher stress response to initial handling and lower feed efficiency, which suggests an involvement of the activation of both the sympatho–adrenal–medullary and hypothalamic–pituitary–adrenal axes on the control of this trait (Cafe et al. 2011). Considerable variability in feeding behaviors within animals was also reported; low feed efficient animals present longer feeding duration and feeding head down time, while high feed efficient animals spend more time being sedentary (Kelly et al. 2010a, b; Chen et al. 2014; McGee et al. 2014).

Considering that efficient animals have lower feed intake to gain the same weight as inefficient animals, it is reasonable to expect that the former spend less time in feeding activities, but why do they eat less? Which mechanisms control the differences in satiety? Appetite is a complex biological process regulated by different signs coming from nutrients, hormones and neural cells that act on the hypothalamus (Sartin et al. 2011). Many studies have focused on understanding the influence of the hormone leptin on feed efficiency because of its role in regulating body weight, feed intake, energy expenditure and even reproduction, inflammation and immunocompetence (Kelly et al. 2010a; Richardson et al. 2004). Although some discordance can be found in the literature, most of the studies show an association of higher leptin levels with lower FE, higher feed intake and higher fat deposition (Richardson et al. 2004; Kelly et al. 2010a; Foote et al. 2016; Nkrumah et al. 2007; Hoque et al. 2009). Polymorphisms within genes related to appetite modulation have also been reported (Santana et al. 2014a, b; Sherman et al. 2008). On the other hand, a few studies trying to understand the complex regulation of FE in the hypothalamus showed no difference in circulating leptin level between high and low FE conditions (Perkins et al. 2014b) or higher expression of leptin in adipose tissue of high FE animals (Perkins et al. 2014a), which indicates that more studies are necessary to understand the modulation of appetite in animals with divergent FE phenotypes. Nevertheless, some important genes were pointed out as regulators of appetite, such as neuropeptide Y (NPY), relaxin-3 (RLN3) and pro-opiomelanocortin (POMC) (Perkins et al. 2014a, b). In another study evaluating differences in hormone expression profile of high and low FE dairy cows, genes regulating adipocytokine signaling pathways and insulin signaling pathways were differentially expressed in two groups (Xi et al. 2015). Interestingly, adipocytokine signaling pathways have leptin as key factor and play an important role in lipid metabolism (Zhao et al. 2013).

It is known that less efficient animals have higher feed intake and that this phenotype can be controlled by differences in metabolism such as satiety control. Reduced liver size has been associated with higher feed efficiency in beef cattle during compensatory gain (Connor et al. 2010). A smaller amount of small intestinal mass was also associated with high feed efficiency, which could indicate decreased nutrient and energy requirements for tissue maintenance, despite the greater mucosal density observed on this phenotype (Montanholi et al. 2013; Meyer et al. 2014). Moreover, lower skeletal muscle protein turnover rate was associated with efficient animals, demonstrating the lower energy requirement for maintenance or, in other words, an energy economy in feed efficient individuals (Castro Bulle et al. 2007).

There are also metabolic consequences of consuming more food which, in turn, impacts on phenotypic measures such as the higher subcutaneous and visceral fat deposition reported in this group of animals (Santana et al. 2012; Nkrumah et al. 2007; Gomes et al. 2012; Basarab et al. 2003; Mader et al. 2009). Higher feed intake is associated with higher proportion of energy needed for maintenance because of the greater energy spent in feeding activities, digestion and tissue metabolism, since an increased size of the digestive organs is expected (Herd et al. 2004; Herd and Arthur 2009).

Tissues of the splanchnic bed (gastrointestinal tract, liver, spleen, pancreas and mesenteric fat depots), together with the associated connective tissue and blood vessels, can be responsible for 35%–60% of the total oxygen consumption of the body (Seal and Reynolds 1993) and this variation seems to be associated with variation in feed intake (Huntington et al. 1988), whereas high feed efficient animals (low feed intake) consume less oxygen (Chaves et al. 2015). Accordingly, lower plasma CO_2 concentrations have been reported in high FE animals, which suggests a decreased oxidation process (Gonano et al. 2014), and in this regard, many studies investigating differences in mitochondrial function with respect to FE can be found. Data in the literature suggest that mitochondrial ADP has greater control of oxidative phosphorylation in the liver of high FE animals (Lancaster et al. 2014) and that increased mitochondrial function in those animals may contribute to improved feed efficiency (Connor et al. 2010). Differences in mitochondrial complex I protein between divergent FE phenotypes have also been reported (Ramos and Kerley 2013). In pigs, differences in mitochondrial function were reported when analyzing muscle (Vincent et al. 2015) and blood transcriptome (Liu et al. 2016), so lower oxidative metabolism was associated with high FE.

In a scenario where low FE animals have a more excitable temperament, higher background energy requirements, higher oxygen consumption and higher CO_2 concentration, it is reasonable to discuss that the associated increase in oxidative metabolism could be coupled with higher energy wastage as heat (Gonano et al. 2014). Indeed, the literature reports warmer thermographs for low FE individuals in thermoneutral environments, which represents a greater amount of heat dissipated by radiation (Montanholi et al. 2010; Montanholi et al. 2009; Archer et al. 1999). Contrarily, in warmer temperature environments, animals activate thermoregulatory functions and, in this case, low FE animals presented colder thermographs, which could indicate less efficient control of body homeostasis (Martello et al. 2016).

As discussed above, many aspects of feed efficiency need to be considered in order to understand the physiological differences related to this trait. In this regard, high-throughput approaches such as sequencing technologies and genome-wide association studies (GWAS), coupled with data integration and network methodologies, have given us some deeper insights into the regulation of this phenotype and hypotheses have been generated. These analyses showed genomic regions and candidate genes associated with feed efficiency of beef cattle (Sherman et al. 2010; Rolf et al. 2012; Lu et al. 2013; Santana et al. 2014a; Oliveira et al. 2014). Figure 1 shows the genome-wide association for ADG, DMI, FCR and RFI in *B. indicus* (Nellore) cattle (Santana et al. 2014a, c). The associated SNPs were related to genes

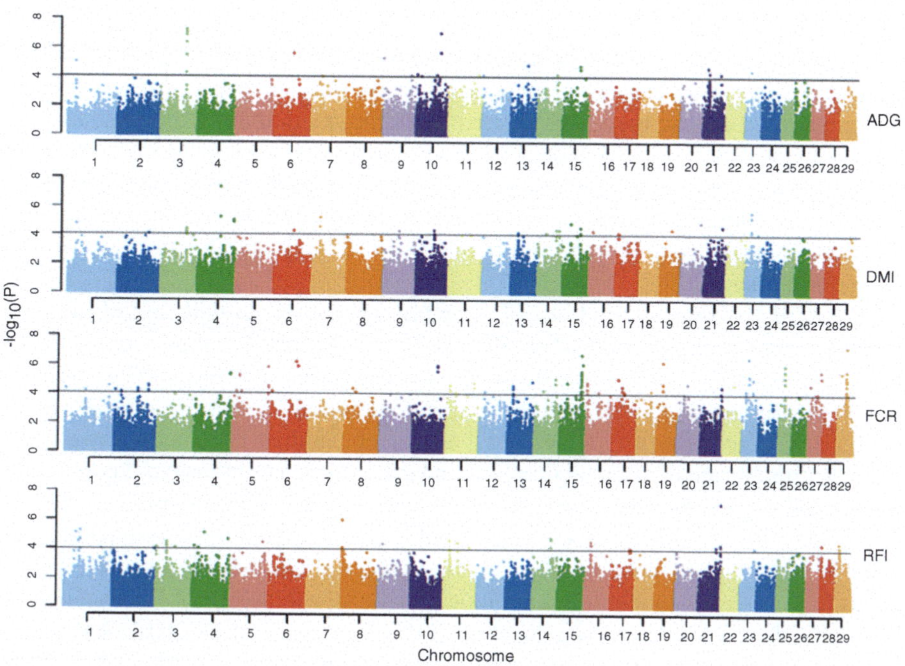

Fig. 1 Manhattan plots of –log (p-value) for average daily gain (ADG), dry matter intake (DMI), feed conversion ratio (FCR) and residual feed intake (RFI). The *horizontal lines* represent the Bonferroni modified significance threshold ($\alpha = 9.27 \times 10^{-5}$)

involved in biological processes like feed intake control, body composition and ion transport.

Karisa, Moore and Plastow (2014) proposed that variations in FE begin with differences in glucose uptake into the cells by the influence of the GHR gene, insulin, creatine-AMPK and leptin on the efficiency of its transport throughout the cytoplasmic membrane. Differences in energy, lipids, steroids and cholesterol metabolism were also identified, which altogether could influence the formation rate of acetyl CoA (Karisa et al. 2014). Indeed, differences in fat deposition and lipid metabolism, with low FE animals showing higher cholesterol levels, were corroborated by our recent study (Alexandre et al. 2015). In ruminants, lipogenesis occurs with limited capacity in the liver and primarily in adipose tissues (Hannun and Obeid 2002). However, with elevated levels of fatty acids, uncoupled NADPH oxidation becomes more probable, generating increased oxygen-derived radicals and hydrogen peroxide in the liver (Knockaert et al. 2011). Higher oxidative stress and expression of antioxidant enzymes in the liver of low FE cattle have already been reported in several studies (Chen et al. 2011; Tizioto et al. 2015; Paradis et al. 2015; Al-Husseini et al. 2014).

Interestingly, in one of our recent papers, network analysis of liver transcriptome from low FE animals also indicated liver inflammation, confirmed by hepatic lesions observed in histopathology analysis and higher level of serum biomarker GGT

(Alexandre et al. 2015). Corroborating our findings, Paradis et al. (2015) found upregulated genes involved with innate immunity in the liver of LFE animals. These authors hypothesized that high FE animals spend less energy to battle systemic inflammation, because they have better hepatic innate immunity response against endotoxins and bacterial products. Based on our results, we conclude that hepatic lesions caused increased liver inflammation because these lesions were probably due to the metabolic stress generated by altered energy/lipid metabolism and/or due to increased bacterial infection from portal blood. Data in the literature support the two possibilities.

The oxidative stress found in LFE animals can also be an outcome of immune response and this is the point where the two previously presented possibilities connect. In pigs, higher intestinal inflammation and neutrophil infiltration biomarkers, together with increased serum endotoxin, were reported for low FE animals (Mani et al. 2013). The authors hypothesized that efficient animals have better capacity to clear, neutralize and detoxify endotoxins and that differences in bacterial population could partially explain the decrease in circulating endotoxins (Mani et al. 2013). Conversely, in cattle, differences in intestinal bacterial population between high and low FE animals have already been reported (Myer et al. 2016). Moreover, differences in genes responsible for xenobiotic metabolism in divergent FE phenotypes have already been shown (Tizioto et al. 2015; Alexandre et al. 2014; Alexandre et al. 2015; Chen et al. 2011). However, we cannot ignore that low FE animals have higher feed intake and this fact can be the cause of liver lesion and the consequent inflammation. The high concentrate (corn and soy) diet provided in feedlot systems is a stressful challenge to the animals which could lead to acidosis and even ruminitis (Owens et al. 1998; Nagaraja and Lechtenberg 2007; Lechtenberg et al. 1988). Animals with higher feed intake also have higher prevalence of liver abscesses caused by bacteria from rumen (Nagaraja and Lechtenberg 2007). It is important to mention here that differences in rumen microbial population have already been pointed out as a contributing factor for FE (Myer et al. 2015).

4 Application of Systems Biology in Feed Efficiency in Beef Cattle

As one can see above, several groups work with different approaches regarding the complex trait of feed efficiency in beef cattle. Studies on behavior to single base polymorphisms in DNA could be found related to FE as well as gene expression from single genes in a given tissue to RNA-seq. Nonetheless, to date, no one has considered a truly systems biology approach with a holistic vision to characterize FE in beef cattle in depth. To perform such important studies, one should consider covering the totality (or as many as possible) of cell types, tissues, organs and the way they interact, i.e. plasma/serum/blood using the most quantitative and in-depth molecular biology techniques to analyze biological macromolecules such as proteins, RNAs and DNA or other molecules such as lipids and metabolites. The "-omics" are necessary to perform a systems biology approach of any trait of interest in any area of biological sciences. In addition, powerful computational tools are necessary to work with and understand the complexity of such huge databases (transcriptome, proteome, lipidome, metabolome, etc.) and propose models of interactions between the molecules from different cell types, tissues and organs (Fig. 2). However, the use of cutting-edge molecular biology tools and advanced bioinformatics does not guarantee good results per se. Clearly, one should be aware that experimental design and sampling are two of the most important factors in a systems biology approach, since this kind of experiments will cost much more than regular experiments. An adequate number of animals, their age, sex, diet, management, etc., are equally important to achieving success.

An example of systems biology application should first consider the definition of the important cells/tissues/organs for a given phenotype. For feed efficiency in beef cattle, one should at least consider the rumen, small intestines, liver, pancreas, adrenals, hypophysis and hypothalamus and plasma or serum. In addition, ruminal content could also be sampled for metagenomics. Next, one can consider using transcriptomic (Fig. 3) and/or proteomic analyses of specific cells and tissues, serum/plasma lipidomics and/or metabolomics coupled with other analytical

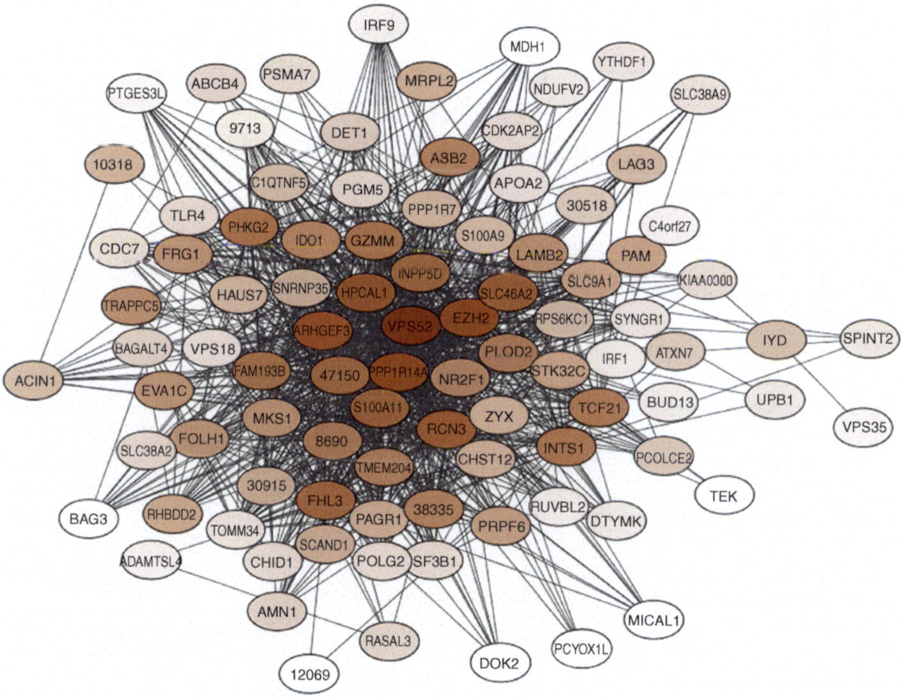

Fig. 2 Example of a model for undirected gene interactions. Genes are represented by nodules and correlations are represented by the edges. Genes represented by the *darkest colors* are central in the network (hub genes) and so, a change in these genes expression would affect greatly the other genes (Data from Alexandre et al. 2015)

analyses such as serum biochemistry and histology. In parallel, global DNA analysis of the same animals should be performed as well or by SNP arrays, whole genome sequencing or the more targeted exome sequencing. The polymorphisms in DNA could be coupled with global gene expression in a given tissue, providing results known as eQTLs (expression quantitative trait loci). The genomic regions associated with specific profile patterns of gene expression identify possible regions of *cis*- and *trans*-regulatory elements in the DNA as promoters, enhancers, silencers and insulators. The sum of these results should be analyzed with care and attention by a multidisciplinary team, since a biased vision of one specific area of knowledge could miss important findings and viewpoints. Remember, the essence of systems biology is the holistic point of view!.

After determining a systems biology model for a trait with the tools mentioned above, it is possible to go to the next level: analysis of the environmental impact on the trait by the regulation of gene expression, a.k.a. epigenetics. However,

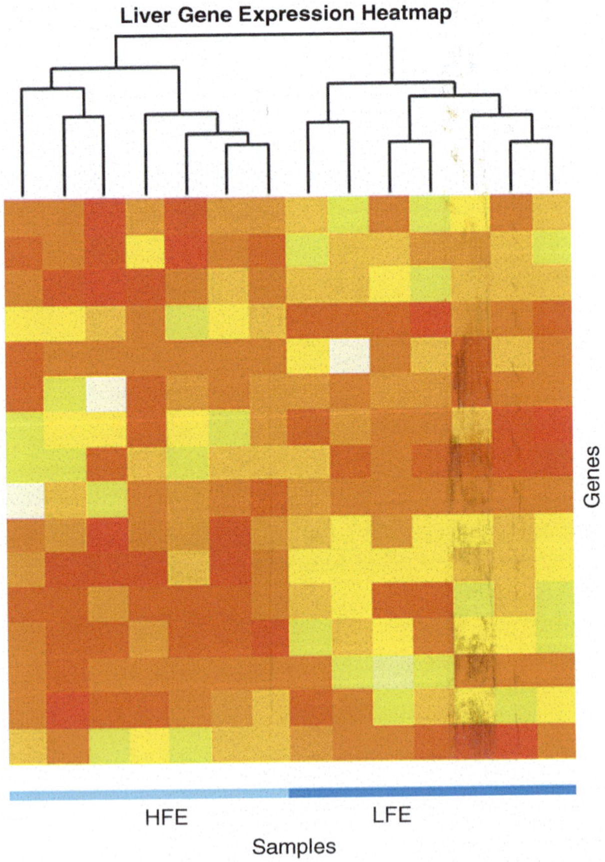

Fig. 3 Heatmap of liver gene expression of the top 16 most varying genes between low feed efficiency (*LFE*) and high feed efficiency (*HFE*). Differences in the expression profile of the two groups are observed by the intensity of the colors representing each gene expression value (Data from Alexandre et al. 2015)

researchers should consider performing this analysis first (and maybe only) in central cell types or tissues for the trait, mostly to understand how the environment interacts with genes by regulating their expression. Understanding the epigenetics of a trait makes possible the search for specific substances that can modulate the trait centrally. In this context, global analysis or gene-specific analysis can be made; the main focus should be on different mechanisms of control of gene expression, such as chromatin and DNA modifications controlling the initiation of mRNA transcription, alternative splicing of mRNAs and its degradation by miRNAs or other non-coding RNAs (ncRNAs). Most recent techniques can evaluate whole DNA epigenetic modifications as the methylome-seq made by bisulfite conversion of methylated cytosines in uracil (RRBS-seq) or by immunoprecipitation of methylated cytosines for DNA microarrays (MeDIP-chip) or next-generation sequencing (MeDIP-seq). It is also possible to characterize the open chromatin regions (regions with potential genes being expressed) by different techniques such as ATAC-seq (assay for transposase-accessible chromatin with high-throughput sequencing), THS-seq (transposome hypersensitive sites sequencing) and DNS-seq (differential nuclease sensitivity sequencing). More recently, methods were designed to evaluate the relationship between multiple epigenetic modifications on the same DNA molecule, for example, NOMe-seq (nucleosome occupancy and methylome sequencing). Another possibility is to determine the sequences of DNA regions associated with regulatory activity by DNase-seq (DNase I hypersensitive sites sequencing) and, more recently, by FAIRE-seq (formaldehyde-assisted isolation of regulatory elements). The epigenomics of specific cells or tissues will provide great evidence on gene expression patterns related to specific phenotypes generating information on possible ways to modulate them.

Alternative splicing of mRNAs in a given sample can be analyzed in theory by RNA-seq but with limitations, since the size of reads can be a problem for correct alignment and using the reference genome makes the detection of trans-splicing almost impossible. An innovative approach involves using nanopore DNA sequencer technology to generate long DNA sequences from cDNAs of specific primer RT reactions. However, this approach is still not high-throughput and gives information on a few genes at the same time, depending on primer design. Global miRNAs and long ncRNAs can also be determined by poly-A selection of mRNAs and RNA-seq or more specifically with commercially available kits for small RNAs and sequencing. This type of work has already been done in cattle divergently selected for residual feed intake (al-Husseini et al. 2015) and in pigs (Jing et al 2015).

Together, all information obtained from experiments analyzed with advanced bioinformatics tools should provide new hypotheses for modulating the phenotype more accurately. This will be possible by breeding or by controlled alterations in the environment (mainly diet and management).

5 Final Remarks and Perspectives

It is important to consider that selection for higher feed efficiency can influence other important economical traits. The differences in body composition and in intermediary metabolism discussed previously can impact reproductive traits. High FE heifers present less fat deposition and later sexual maturity, which in turn results in calving later in the calving season than low FE herdmates (Randel and Welsh 2013; Shaffer et al. 2011). Furthermore, feed efficient bulls also have features of delayed sexual maturity, for instance, decreased progressive motility and higher abundance of tail abnormalities of the sperm (Fontoura et al. 2016). At present, in Brazil, selection for early pregnancy of Nellore heifers is considered a very important selection criterion. However, if that conflicts with feed efficiency, a broader approach should be strongly recommended. Early pregnancy heifers mean more calves/cow, fewer cows to produce the same amount of beef and less gas emission. But if more efficient cattle are not so precious, how one will be able to balance efficiency versus production is an open question. Definitely, more studies are necessary to understand FE biology in order to help us select efficient animals without compromising other important production traits. This will certainly be related to the application of systems biology on this important trait.

Conclusions

- Feed efficiency in beef cattle is a very important and complex biological trait which is controlled by genetic and epigenetic effects.
- New molecular biology techniques and bioinformatics tools are already being used on feed efficiency experiments.
- The use of systems biology will definitely speed up the gain of knowledge regarding the regulation of feed efficiency.

References

Aggrey SE, Rekaya R (2013) Dissection of Koch's residual feed intake: implications for selection. Poult Sci 92:2600–2605. doi:10.3382/ps.2013-03302

Alexandre P a et al (2014) Bovine NR1I3 gene polymorphisms and its association with feed efficiency traits in Nellore cattle. Meta Gene 2:206–217, Available at: http://linkinghub.elsevier.com/retrieve/pii/S2214540014000048. Accessed 15 Aug 2014

Alexandre PA et al (2015) Liver transcriptomic networks reveal main biological processes associated with feed efficiency in beef cattle. BMC Genomics 16(1):1073. Available at: http://www.biomedcentral.com/1471-2164/16/1073

Al-Husseini W et al (2014) Expression of candidate genes for residual feed intake in Angus cattle. Anim Genetics 45(1):12–19. Available at: http://www.ncbi.nlm.nih.gov/pubmed/24134470. Accessed 1 Feb 2015

Al-Husseini W et al (2015) Characterization and profiling of liver microRNAs by RNA-sequencing in cattle divergently selected for residual feed intake. Asian-Australasian J Anim Sci. Available at: http://www.ncbi.nlm.nih.gov/pubmed/26954124. Accessed 10 May 2016

Archer JA et al (1999) Potential for selection to improve efficiency of feed use in beef cattle: a review. Aust J Agr Res 50(2):147–162, Available at: http://www.publish.csiro.au/paper/A98075.htm. Accessed 29 Feb 2016

Arthur P, Renand G, Krauss D (2001a) Genetic parameters for growth and feed efficiency in weaner versus yearling Charolais bulls. Aust J Exp Agric 52:471–476

Arthur PF, Archer JA, Johnston DJ, Herd RM, Richardson EC, Parnell PF (2001b) Genetic and phenotypic variance and covariance components for feed intake, feed efficiency, and other postweaning traits in Angus cattle. J Anim Sci 79:2805–2811

Arthur PF, Herd RM, Wilkins JF, Archer JA (2005) Maternal productivity of Angus cows divergently selected for post-weaning residual feed intake. Aust J Exp Agric 45:985. doi:10.1071/EA05052

Atchley WR, Anderson D (1978) Ratios and the statistical analysis of biological data. Syst Zool 27:71. doi:10.2307/2412816

Barwick S a, Wolcott ML, Johnston DJ, Burrow HM, Sullivan MT (2009) Genetics of steer daily and residual feed intake in two tropical beef genotypes, and relationships among intake, body composition, growth and other post-weaning measures. Animal Prod Sci 49:351. doi:10.1071/EA08249

Basarab J a et al (2003) Residual feed intake and body composition in young growing cattle. Can J Anim Sci 83(2):189–204, Available at: http://pubs.aic.ca/doi/abs/10.4141/A02-065

Basarab JA, McCartney D, Okine EK, Baron VS (2007) Relationships between progeny residual feed intake and dam productivity traits. Can J Anim Sci 87:489–502. doi:10.4141/CJAS07026

Berry DP, Crowley JJ (2012) Residual intake and body weight gain: a new measure of efficiency in growing cattle. J Anim Sci 90:109–115. doi:10.2527/jas.2011-4245

Cafe LM et al (2011) Temperament and hypothalamic-pituitary-adrenal axis function are related and combine to affect growth, efficiency, carcass, and meat quality traits in Brahman steers. Domest Anim Endocrinol 40(4):230–240, Available at: http://www.sciencedirect.com/science/article/pii/S0739724011000063. Accessed 21 Feb 2016

Canovas A, Reverter A, DeAtley KL et al (2014) Multi-tissue omics analyses reveal molecular regulatory networks for puberty in composite beef. PLoS One 9(7), e102551. doi:10.1371/journal.pone.0102551Cattle

Castro Bulle FCP et al (2007) Growth, carcass quality, and protein and energy metabolism in beef cattle with different growth potentials and residual feed intakes. J Anim Sci 85(4):928–936. Available at: http://www.ncbi.nlm.nih.gov/pubmed/17178805. Accessed 1 Mar 2016

Chaves AS et al (2015) Relationship of efficiency indices with performance, heart rate, oxygen consumption, blood parameters, and estimated heat production in Nellore steers. J Anim Sci 93(10):5036–5046. Available at: http://www.ncbi.nlm.nih.gov/pubmed/26523596. Accessed 28 Feb 2016

Chen Y et al (2011) Global gene expression profiling reveals genes expressed differentially in cattle with high and low residual feed intake. Anim Genetics 42(5):475–490. Available at: http://www.ncbi.nlm.nih.gov/pubmed/21906099. Accessed 14 Apr 2012

Chen L et al (2014) Phenotypic and genetic relationships of feeding behavior with feed intake, growth performance, feed efficiency, and carcass merit traits in Angus and Charolais steers. J Anim Sci 92(3):974–983. Available at: http://www.ncbi.nlm.nih.gov/pubmed/24492561. Accessed 21 Feb 2016

Connor EE et al (2010) Enhanced mitochondrial complex gene function and reduced liver size may mediate improved feed efficiency of beef cattle during compensatory growth. Funct Integrat Genomics 10(1):39–51. Available at: http://www.ncbi.nlm.nih.gov/pubmed/19777276. Accessed 28 Feb 2016

Crews DH (2005) Genetics of efficient feed utilization and national cattle evaluation: a review. Genet Mol Res 4:152–165

Ferraz JBS, Felício PE (2010) Production systems – an example from Brazil. Meat Sci 84:238–243

Fitzhugh HA Jr, Taylor CS (1971) Genetic analysis of degree of maturity. J Anim Sci 33:717–725

Fontoura ABP et al (2016) Associations between feed efficiency, sexual maturity and fertility-related measures in young beef bulls. Animal Int J Anim Biosci 10(1):96–105. Available at: http://www.ncbi.nlm.nih.gov/pubmed/26351012. Accessed 2 Mar 2016

Foote AP et al (2016) Leptin concentrations in finishing beef steers and heifers and their association with dry matter intake, average daily gain, feed efficiency, and body composition. Domestic Anim Endocrinol 55:136–141. Available at: http://www.ncbi.nlm.nih.gov/pubmed/26851619. Accessed 23 Feb 2016

Francisco CL et al (2015) Impacts of temperament on Nellore cattle: physiological responses, feedlot performance, and carcass characteristics. J Anim Sci 93(11):5419. Available at: http://www.ncbi.nlm.nih.gov/pubmed/26641061. Accessed 21 Feb 2016

Gomes RC et al (2012) Feedlot performance, feed efficiency reranking, carcass traits, body composition, energy requirements, meat quality and calpain system activity in Nellore steers with low and high residual feed intake. Livest Sci 150(1-3):265–273, Available at: http://www.sciencedirect.com/science/article/pii/S1871141312003587. Accessed 19 Feb 2014

Gonano C et al (2014) The relationship between feed efficiency and the circadian profile of blood plasma analytes measured in beef heifers at different physiological stages. Anim 8(10):1684–1698. Available at: http://www.ncbi.nlm.nih.gov/pubmed/24923431. Accessed 28 Feb 2016

Grion AL, Mercadante MEZ, Cyrillo JNSG, Bonilha SFM, Magnani E, Branco RH (2014) Selection for feed efficiency traits and correlated genetic responses in feed intake and weight gain of Nellore cattle. J Anim Sci 92:955–965. doi:10.2527/jas.2013-6682

Gunsett FC (1984) Linear index selection to improve traits defined as ratios. J Anim Sci 59:1185–1193. doi:10.2134/jas1984.5951185x

Hannun YA, Obeid LM (2002) The Ceramide-centric universe of lipid-mediated cell regulation: stress encounters of the lipid kind. J Biol Chem 277(29):25847–25850. Available at: http://www.ncbi.nlm.nih.gov/pubmed/12011103. Accessed 3 July 2015

Hegarty RS, Goopy JP, Herd RM, McCorkell B (2007) Cattle selected for lower residual feed intake have reduced daily methane production. J Anim Sci 85:1479–1486. doi:10.2527/jas.2006-236

Herd RM, Arthur PF (2009) Physiological basis for residual feed intake. J Anim Sci 87(14 Suppl):E64–E71, Available at: http://jas.fass.org/cgi/content/abstract/87/14_suppl/E64. Accessed 28 Mar 2012

Herd RM, Oddy VH, Richardson EC (2004) Biological basis for variation in residual feed intake in beef cattle. 1. Review of potential mechanisms. Aust J Exp Agric 44(5):423. Available at: http://www.publish.csiro.au/view/journals/dsp_journal_fulltext.cfm?nid=72&f=EA02220. Accessed 18 Apr 2012

Hoque MA, Katoh K, Suzuki K (2009) Genetic associations of residual feed intake with serum insulin-like growth factor-I and leptin concentrations, meat quality, and carcass cross sectional fat area ratios in Duroc pigs. J Anim Sci 87(10):3069–3075, Available at: https://www.animalsciencepublications.org/publications/jas/abstracts/87/10/3069. Accessed 23 Feb 2016

Huntington GB et al (1988) Net absorption and oxygen consumption by Holstein steers fed alfalfa or orchardgrass silage at two equalized intakes. J Anim Sci 66(5):1292–1302. Available at: http://www.ncbi.nlm.nih.gov/pubmed/3397352. Accessed 13 Jan 2015

Jing L et al (2015) Transcriptome analysis of mRNA and miRNA in skeletal muscle indicates an important network for differential residual feed intake in pigs. Sci Rep 5:11953, Available at: http://www.pubmedcentral.nih.gov/articlerender.fcgi?artid=4493709&tool=pmcentrez&rendertype=abstract. Accessed 6 Apr 2016

Kadarmideen HN (2014) Genomics to systems biology in animal and veterinary sciences: Progress, lessons and opportunities. Livestock Sci 166(2014):232–248. doi:10.1016/j.livsci.2014.04.028

Karisa B, Moore S, Plastow G (2014) Analysis of biological networks and biological pathways associated with residual feed intake in beef cattle. Anim Sci J = Nihon chikusan Gakkaihō 85(4):374–387. Available at: http://www.ncbi.nlm.nih.gov/pubmed/24373146. Accessed 1 Feb 2015

Kelly AK, McGee M, Crews DH, Fahey AG et al (2010a) Effect of divergence in residual feed intake on feeding behavior, blood metabolic variables, and body composition traits in growing beef heifers. J Anim Sci 88(1):109–123. Available at: http://www.ncbi.nlm.nih.gov/pubmed/19820067. Accessed 21 Feb 2016

Kelly AK, McGee M, Crews DH, Sweeney T et al (2010b) Repeatability of feed efficiency, carcass ultrasound, feeding behavior, and blood metabolic variables in finishing heifers divergently selected for residual feed intake. J Anim Sci 88(10):3214–3225. Available at: http://www.ncbi.nlm.nih.gov/pubmed/20525931. Accessed 21 Feb 2016

Kennedy BW, van der Werf JH, Meuwissen TH (1993) Genetic and statistical properties of residual feed intake. J Anim Sci 71:3239–3250

KLEIBER M (1947) Body size and metabolic rate. Physiol Rev 27:511–541

Knockaert L, Fromenty B, Robin M-A (2011) Mechanisms of mitochondrial targeting of cytochrome P450 2E1: physiopathological role in liver injury and obesity. FEBS J 278(22):4252–4260. Available at: http://www.ncbi.nlm.nih.gov/pubmed/21929725. Accessed 2 Feb 2015

Koch RM, Swiger LA, Chambers D, Gregory KE (1963) Efficiency of feed use in beef cattle. J Anim Sci 22:486–494

Lancaster PA et al (2014) Relationships between residual feed intake and hepatic mitochondrial function in growing beef cattle. J Anim Sci 92(7):3134–3141. Available at: http://www.ncbi.nlm.nih.gov/pubmed/24894006. Accessed 29 Feb 2016

Lechtenberg KF et al (1988) Bacteriologic and histologic studies of hepatic abscesses in cattle. Am J Veterinary Res 49(1):58–62. Available at: http://www.ncbi.nlm.nih.gov/pubmed/3354968. Accessed 1 Feb 2015

Liu H et al (2016) Post-weaning blood transcriptomic differences between Yorkshire pigs divergently selected for residual feed intake. BMC Genomics 17(1):73. Available at: http://www.pubmedcentral.nih.gov/articlerender.fcgi?artid=4724083&tool=pmcentrez&rendertype=abstract. Accessed 1 Feb 2016

Lobato JFP et al (2014) Brazilian beef produced on pastures: sustainable and healthy. Meat Sci 98:336–345

Lu D, Miller S, Sargolzaei M, Kelly M, Vander Voort G, Caldwell T, Wang Z, Plastow G, Moore S (2013) Genome-wide association analyses for growth and feed efficiency traits in beef cattle. J Anim Sci 91:3612–3633. doi:10.2527/jas.2012-5716

Mader CJ et al (2009) Relationships among measures of growth performance and efficiency with carcass traits, visceral organ mass, and pancreatic digestive enzymes in feedlot cattle. J Anim Sci 87(4):1548–1557. Available at: http://www.ncbi.nlm.nih.gov/pubmed/18952722. Accessed 30 Jan 2015

Mani V et al (2013) Intestinal integrity, endotoxin transport and detoxification in pigs divergently selected for residual feed intake. J Anim Sci 91(5):2141–2150. Available at: http://www.ncbi.nlm.nih.gov/pubmed/23463550. Accessed 1 Mar 2016

Martello LS et al (2016) Infrared thermography as a tool to evaluate body surface temperature and its relationship with feed efficiency in Bos indicus cattle in tropical conditions. Int J Biometeorol 60(1):173–181. Available at: http://www.ncbi.nlm.nih.gov/pubmed/260703691 Accessed 29 Feb 2016

McGee M et al (2014) Relationships of feeding behaviors with average daily gain, dry matter intake, and residual feed intake in Red Angus-sired cattle. J Anim Sci 92(11):5214–5221. Available at: http://www.ncbi.nlm.nih.gov/pubmed/25349363. Accessed 21 Feb 2016

Meyer PM, Rodrigues PHM (2014) Progress in the Brazilian cattle industry: an analysis of the Agricultural Censuses database. Anim Product Sci 54:1338–1344

Meyer AM et al (2014) Small intestinal growth measures are correlated with feed efficiency in market weight cattle, despite minimal effects of maternal nutrition during early to midgestation. J Anim Sci 92(9):3855–3867. Available at: http://www.ncbi.nlm.nih.gov/pubmed/25057033. Accessed 28 Feb 2016

Millen DD, Arrigoni MDB (2013) Drivers of change in animal protein production systems: Changes from'traditional' to 'modern' beef cattle production systems in Brazil. Anim Front 3(3):56–60

Montaldo HH, Casas E, Ferraz JBS, Vega-Murillo VE, Roman-Ponce SI (2012) Opportunities and challenges from the use of genomic selection for beef cattle breeding in Latin America. Anim Front Rev Magazine Animal Agric 2:23–29

Montanholi YR et al (2009) On the determination of residual feed intake and associations of infrared thermography with efficiency and ultrasound traits in beef bulls. Livest Sci 125(1):22–30, Available at: http://www.sciencedirect.com/science/article/pii/S1871141309000766. Accessed 29 Feb 2016

Montanholi YR et al (2010) Assessing feed efficiency in beef steers through feeding behavior, infrared thermography and glucocorticoids. Anim Int J Animal Biosci 4(5):692–701. Available at: http://www.ncbi.nlm.nih.gov/pubmed/22444121. Accessed 29 Feb 2016

Montanholi Y et al (2013) Small intestine histomorphometry of beef cattle with divergent feed efficiency. Acta veterinaria Scandinavica 55:9. Available at: http://www.pubmedcentral.nih.gov/articlerender.fcgi?artid=3598877&tool=pmcentrez&rendertype=abstract. Accessed 28 Feb 2016

Myer PR et al (2015) Rumen microbiome from steers differing in feed efficiency. PloS One 10(6):e0129174. Available at: http://www.pubmedcentral.nih.gov/articlerender.fcgi?artid=4451142&tool=pmcentrez&rendertype=abstract. Accessed 1 Mar 2016

Myer PR et al (2016) Microbial community profiles of the jejunum from steers differing in feed efficiency. J Anim Sci 94(1):327–338. Available at: http://www.ncbi.nlm.nih.gov/pubmed/26812338. Accessed 27 Jan 2016

Nagaraja TG, Lechtenberg KF (2007) Liver abscesses in feedlot cattle. The Veterinary clinics of North America. Food Anim Pract 23(2):351–369, ix. Available at: http://www.ncbi.nlm.nih.gov/pubmed/17606156. Accessed 31 Jan 2015

Nkrumah J, Okine E, Mathison G, Schmid K, Li C, Basarab JA, Price MA, Wang Z, Moore SS (2006) Relationships of feedlot feed efficiency, performance, and feeding behavior with metabolic rate, methane production, and energy partitioning in beef cattle. J Anim 84:145–153

Nkrumah JD et al (2007) Genetic and phenotypic relationships of serum leptin concentration with performance, efficiency of gain, and carcass merit of feedlot cattle. J Anim Sci 85(9):2147–2155. Available at: http://www.ncbi.nlm.nih.gov/pubmed/17468416. Accessed 23 Feb 2016

Oliveira PSN, Cesar ASM, Nascimento ML, Chaves AS, Tizioto PC, Tullio RR, Lanna DPD, Rosa AN, Sonstegard TS, Mourao GB, Reecy JM, Garrick DJ, Mudadu MA, Coutinho LL, Regitano LCA (2014) Identification of genomic regions associated with feed efficiency in Nelore cattle. BMC Genet 15:100. doi:10.1186/s12863-014-0100-0

Owens FN et al (1998) Acidosis in cattle: a review. J Anim Sci 76(1):275–286. Available at: http://www.ncbi.nlm.nih.gov/pubmed/9464909. Accessed 1 Feb 2015

Paradis F et al (2015) Transcriptomic analysis by RNA sequencing reveals that hepatic interferon-induced genes may be associated with feed efficiency in beef heifers. J Anim Sci 93(7):3331–3341. Available at: http://www.ncbi.nlm.nih.gov/pubmed/26440002. Accessed 8 Nov 2015

Perkins SD, Key CN, Garrett CF et al (2014) Residual feed intake studies in Angus-sired cattle reveal a potential role for hypothalamic gene expression in regulating feed efficiency. J Anim Sci 92(2):549–560. Available at: http://www.ncbi.nlm.nih.gov/pubmed/24398827. Accessed 23 Feb 2016

Perkins SD, Key CN, Marvin MN et al (2014) Effect of residual feed intake on hypothalamic gene expression and meat quality in Angus-sired cattle grown during the hot season. J Anim Sci 92(4):1451–1461. Available at: http://www.ncbi.nlm.nih.gov/pubmed/24663166. Accessed 23 Feb 2016

Ramos MH, Kerley MS (2013) Mitochondrial complex I protein differs among residual feed intake phenotype in beef cattle. J Anim Sci 91(7):3299–3304. Available at: http://www.ncbi.nlm.nih.gov/pubmed/23798519. Accessed 29 Feb 2016

Randel RD, Welsh TH (2013) Joint Alpharma-Beef Species Symposium: interactions of feed efficiency with beef heifer reproductive development. J Anim Sci 91(3):1323–1328. Available at: http://www.ncbi.nlm.nih.gov/pubmed/23048157. Accessed 2 Mar 2016

Richardson EC et al (2004) Metabolic differences in Angus steers divergently selected for residual feed intake. Aust J Exp Agric 44(5):441. Available at: http://www.publish.csiro.au/view/journals/dsp_journal_fulltext.cfm?nid=72&f=EA02219. Accessed 18 Apr 2012

Rolf MM, Taylor JF, Schnabel RD, McKay SD, McClure MC, Northcutt SL, Kerley MS, Weaber RL (2012) Genome-wide association analysis for feed efficiency in Angus cattle. Anim Genet 43:367–374. doi:10.1111/j.1365-2052.2011.02273.x

Rolfe KM, Snelling WM, Nielsen MK, Freetly HC, Ferrell CL, Jenkins TG (2011) Genetic and phenotypic parameter estimates for feed intake and other traits in growing beef cattle, and opportunities for selection. J Anim Sci 89:3452–3459. doi:10.2527/jas.2011-3961

Santana MHA et al (2012) Feed efficiency and its correlations with carcass traits measured by ultrasound in Nellore bulls. Livest Sci 145(1-3):252–257, Available at: http://linkinghub.elsevier.com/retrieve/pii/S1871141312000686. Accessed 13 Jan 2015

Santana MH et al (2014a) Genome-wide association analysis of feed intake and residual feed intake in Nellore cattle. BMC Genetics 15(1):21. Available at: http://www.ncbi.nlm.nih.gov/pubmed/24517472. Accessed 19 Feb 2014

Santana MH et al (2014b) Single nucleotide polymorphisms in genes linked to ion transport and regulation of appetite and their associations with weight gain, feed efficiency and intake of Nellore cattle. Livestock Sci 165(1):33–36. Available at: http://dx.doi.org/10.1016/j.livsci.2014.04.004

Santana MHA et al (2014c) Genome-wide association study for feedlot average daily gain in Nellore cattle (Bos indicus). J Anim Breed Genet 131:210–216. doi:10.1111/jbg.12084

Sartin JL, Whitlock BK, Daniel JA (2011) Triennial Growth Symposium: neural regulation of feed intake: modification by hormones, fasting, and disease. J Anim Sci 89(7):1991–2003. Available at: http://www.ncbi.nlm.nih.gov/pubmed/21148776. Accessed 23 Feb 2016

Seal CJ, Reynolds CK (1993) Nutritional implications of gastrointestinal and liver metabolism in ruminants. Nutrition Res Rev 6(1):185–208. Available at: http://www.ncbi.nlm.nih.gov/pubmed/19094308. Accessed 28 Feb 2016

Shaffer KS et al (2011) Residual feed intake, body composition, and fertility in yearling beef heifers. J Anim Sci 89(4):1028–1034. Available at: http://www.ncbi.nlm.nih.gov/pubmed/21112981. Accessed 2 Mar 2016

Sherman EL et al (2008) Polymorphisms and haplotypes in the bovine neuropeptide Y, growth hormone receptor, ghrelin, insulin-like growth factor 2, and uncoupling proteins 2 and 3 genes and their associations with measures of growth, performance, feed efficiency, and carcass meri. J Anim Sci 86(1):1–16. Available at: http://www.ncbi.nlm.nih.gov/pubmed/17785604. Accessed 24 Feb 2016

Sherman EL, Nkrumah JD, Moore SS (2010) Whole genome single nucleotide polymorphism associations with feed intake and feed efficiency in beef cattle. J Anim Sci 88:16–22. doi:10.2527/jas.2008-1759

Tizioto PC et al (2015) Global liver gene expression differences in Nelore steers with divergent residual feed intake phenotypes. BMC Genomics 16(1), Available at: http://www.biomedcentral.com/1471-2164/16/242

van der Werf JHJ (2004) Is it useful to define residual feed intake as a trait in animal breeding programs? Aust J Exp Agric 44:405. doi:10.1071/EA02105

Vincent A et al (2015) Divergent selection for residual feed intake affects the transcriptomic and proteomic profiles of pig skeletal muscle. J Anim Sci 93(6):2745–2758. Available at: http://www.ncbi.nlm.nih.gov/pubmed/26115262. Accessed 29 Feb 2016

Widmann P, Reverter A, Fortes MRS et al (2013) A systems biology approach using metabolomic data reveals genes and pathways interacting to modulate divergent growth in cattle. BMC Genomics 14:798

Xi YM et al (2015) Gene expression profiling of hormonal regulation related to the residual feed intake of Holstein cattle. Biochem Biophys Res Commun 465(1):19–25, Available at: http://www.sciencedirect.com/science/article/pii/S0006291X15303223. Accessed 23 Feb 2016

Zhao M, Li X, Qu H (2013) EDdb: a web resource for eating disorder and its application to identify an extended adipocytokine signaling pathway related to eating disorder. Science China. Life Sci 56(12):1086–1096. Available at: http://www.ncbi.nlm.nih.gov/pubmed/24302289. Accessed 23 Feb 2016

Nutritional Systems Biology to Elucidate Adaptations in Lactation Physiology of Dairy Cows

Mario Vailati-Riboni, Ahmed Elolimy, and Juan J. Loor

Abstract

A fuller understanding of the complexity of physiologic and metabolic adaptations experienced by the modern high-producing dairy cow during the transition into lactation unavoidably requires application of systems biology, i.e. a way to systematically study the biological interactions within the cow using a method of integration instead of reduction. The use of high-throughput technologies, i.e. "omics," along with bioinformatics are ideal for uncovering pathways, regulatory networks, and structural organization within and between tissues (e.g. adipose and liver, skeletal muscle and adipose, gut microorganisms and epithelia). The integration of this information results in a more holistic appreciation of how the cow functions, particularly when used within a framework encompassing nutrition as a tool for optimizing the ability to adapt to lactation without compromising health. This chapter first outlines the current state of biological knowledge on five key areas identified as crucial for achieving marked gains in productivity. After a brief description of high-throughput technologies, we discuss breakthroughs in knowledge at the tissue, cell, and rumen level. Major topics include regulation of feed intake, immune function, fat deposition, and the rumen microbiota. The goal is to provide specific examples of how genome-enabled approaches have been used to advance our understanding of tissue and cell function, and microbiota adaptations to dietary changes during the transition into lactation. The available research illustrates how a more widespread application of systems biology in ruminant nutrition will, in the medium-to-long-term, enable scientists to design functional diets that enhance dairy cow productivity and health based on exploiting the plasticity of the rumen microbial ecosystem along with the cow's full genetic potential.

M. Vailati-Riboni • A. Elolimy • J.J. Loor (✉)
Department of Animal Sciences and Division of Nutritional Sciences, University of Illinois, Urbana, IL 61801, USA
e-mail: jloor@illinois.edu

H.N. Kadarmideen (ed.), *Systems Biology in Animal Production and Health, Vol. 2*,
DOI 10.1007/978-3-319-43332-5_5

1 The Nutrition–Physiology–Management Relationship in Modern Dairy Cattle

It is well-accepted among dairy nutritionists and physiologists that the period around parturition ("peripartum period") and the first 3 months of lactation are the most challenging in the life cycle of dairy cattle. Achieving homeostasis during the peripartum period and early lactation represents a monumental task in modern high-producing dairy cows (Drackley et al. 2006). It is for this reason that the mechanisms underlying metabolic/physiologic adaptations in keys organs such as liver, adipose, and skeletal muscle during this physiological stage remain active areas of research (Drackley et al. 2006; Loor et al. 2013b; Roche et al. 2013a). There are several excellent reviews published in the last 5 years on the state of knowledge regarding the relationship among nutrition, physiology, and management in peripartum dairy cows (Loor et al. 2013b; Roche et al. 2013a; Bradford et al. 2015; Sordillo 2016).

The main goal of this brief introduction is not to recapitulate the information already available but rather to summarize, in general, the state of knowledge in areas that were proposed by Drackley (1999) to be central for advancing our understanding of the linkage among nutrition, management, and physiology of the dairy cow during the transition from pregnancy into lactation. We also provide some thoughts on proposals that may be a bit controversial at the present time, particularly because of lack of enough data. This section is meant to serve as the basis for a brief presentation and discussion in the other sections of this chapter of examples from work encompassing systems biology concepts related to nutrition and management of dairy cows.

The following are some of the key areas in which additional knowledge proposed by Drackley (1999) would provide the "largest gains in productivity and profitability in the next decade":

1.1 Control of Feed Intake

The first review, to our knowledge, of the potential linkage between metabolism and intake regulation was published in 2000 (Ingvartsen and Andersen 2000). It was proposed then that signals such as "nutrients, metabolites, reproductive hormones, stress hormones, leptin, insulin, gut peptides, cytokines, and neuropeptides such as neuropeptide Y, galanin, and corticotrophin-releasing factor" likely play an "equally important role in intake regulation." A recent review by Allen and Piantoni (2013), however, concluded that control of feed intake during the transition into lactation ("peripartum period") is likely caused by "signals" produced during the oxidation of fuels in the liver. They proposed that "continuous supply of NEFA to the liver during the lipolytic state at this time likely suppresses feed intake, as they are oxidized." Although these authors present some persuasive arguments, we believe that additional mechanistic studies are required to fully test this overarching hypothesis. For instance, additional data on hepatic concentrations of acetyl-CoA across a variety of diets would help establish the degree of association of this end-product of oxidative metabolism and feed intake. The neural control of feed intake also would have to be addressed in a more mechanistic fashion to fully understand what role (if any) peripheral signals discussed by Ingvartsen and Andersen (2000) have on the control of feed intake.

1.2 Quantification of Nutrient Supply in the Face of a Rapid Change in Rate of Feed Intake and Gut Capacity

At least for the liver, a major breakthrough in this area was achieved by Reynolds et al. (2003) who meticulously determined splanchnic metabolism in dairy cows during the peripartum period and early lactation. For instance, it was determined that after parturition the rate of liver metabolism nearly doubles and is a key factor driving the increases in milk production and feed intake. The increase in feed intake was the main determinant of changes in splanchnic metabolism observed during the transition into lactation (Reynolds et al. 2003). Measured contributions of propionate, lactate, alanine, and glycerol to hepatic glucose production were measured for the first time. Recent studies have utilized similar approaches to determine the quantitative aspects of amino acid metabolism not only in splanchnic tissues but also whole-body and in the mammary gland (e.g. Larsen et al. 2015). In addition, a recent review (Zebeli et al. 2015) underscored the role of proper dietary and feeding management of peripartum cows in the context of optimizing the cow's adaptations to lactation. The proper development of the rumen epithelium and other sections of the gastrointestinal tract during the peripartum period clearly could have an impact on allowing the cow metabolism to adapt to the onset of lactation.

1.3 Interactions among Nutrition, Metabolism, and the Immune System

These linkages have received increased attention during the last 10 years, and the state of knowledge has been discussed in at least 5 reviews since 2009 (Sordillo et al. 2009; Bertoni and Trevisi 2013; Ingvartsen and Moyes 2013; Sordillo and Raphael 2013; Sordillo 2016). There is consensus that nutritional management around parturition and early lactation not only alters metabolism, but contributes to proper functioning of the immune system. Overall, it is widely accepted that the immune system of peripartum dairy cows is "dysfunctional," with a number of important functions being substantially reduced not only in neutrophils (e.g. chemotaxis, phagocytosis) but also in monocytes (TNF production) and lymphocytes (total numbers, IFN-γ) (Moyes 2015). Novel beneficial roles for essential nutrients such as methionine in the context of immune function and oxidative stress around parturition have emerged (Osorio et al. 2013b, 2014a, b). Despite substantial progress in this area of research, there is still a gap of knowledge on the regulatory mechanisms whereby specific nutrients, e.g. vitamins, microminerals, essential amino acids, and essential fatty acids, elicit positive effects on the immune and metabolic function of the cow.

1.4 Metabolic Regulation in, and Interactions Among Liver, Adipose, Muscle, and the Digestive Tract

The first comprehensive dataset addressing metabolic regulation in a tissue of dairy cows during the peripartum period were generated by McNamara's laboratory on adipose tissue (McNamara 1991, 1997). The work of Reynolds et al. (2003) provided the most complete dataset of hepatic metabolic regulation during the peripartum period and early lactation. Those original studies have led other scientists to focus on understanding better specific pathways such as gluconeogenesis from the standpoint of carbohydrate (e.g. Aschenbach et al. 2010) or amino acid (Larsen and Kristensen 2013) nutrition. Others in the 1980s had addressed the endocrine regulation of metabolism during lactation (Collier et al. 1984). Despite the vast amount of knowledge on endocrine regulation of metabolism, recent research indicates that there are additional factors, e.g. serotonin, that potentially play a role in coordinating metabolic adaptations to lactation in the cow (Laporta et al. 2015).

1.5 "Body Condition" at Parturition and Its Relationship with Nutrition in the Context of Metabolic Responses to Onset of Lactation

The body condition score (BCS) of a dairy cow is an assessment of the proportion of body fat that it possesses, and it is recognized by animal scientists and producers as being an important factor in dairy cattle management (Roche et al. 2009). It is now widely recognized that for many production and health variables, there is no linear relationship with BCS, i.e. lower BCS at calving is associated with lower production and impaired reproductive capacity, while higher BCS is associated with a reduction in feed intake and milk production and a higher risk of metabolic disorders (Roche et al. 2013b). Work to date clearly underscores that genetics of the cow, nutrition, and management are key factors interacting with BCS to determine the risk of health disorders (Roche et al. 2013b).

Clearly, there has been a substantial amount of progress generated on the linkages among nutrition, physiology, and management of dairy cows during the critical peripartum period. However, there are still important gaps in knowledge at an organ level and more importantly at the systems level. The remaining sections of this chapter introduce important technologies that already have been used to better understand regulatory mechanisms at the gene, transcript, protein, and metabolite levels. Figure 1 depicts the conceptual framework for the application of systems biology approaches to better understand the underlying mechanisms associated with successful and unsuccessful physiologic adaptations of lactation.

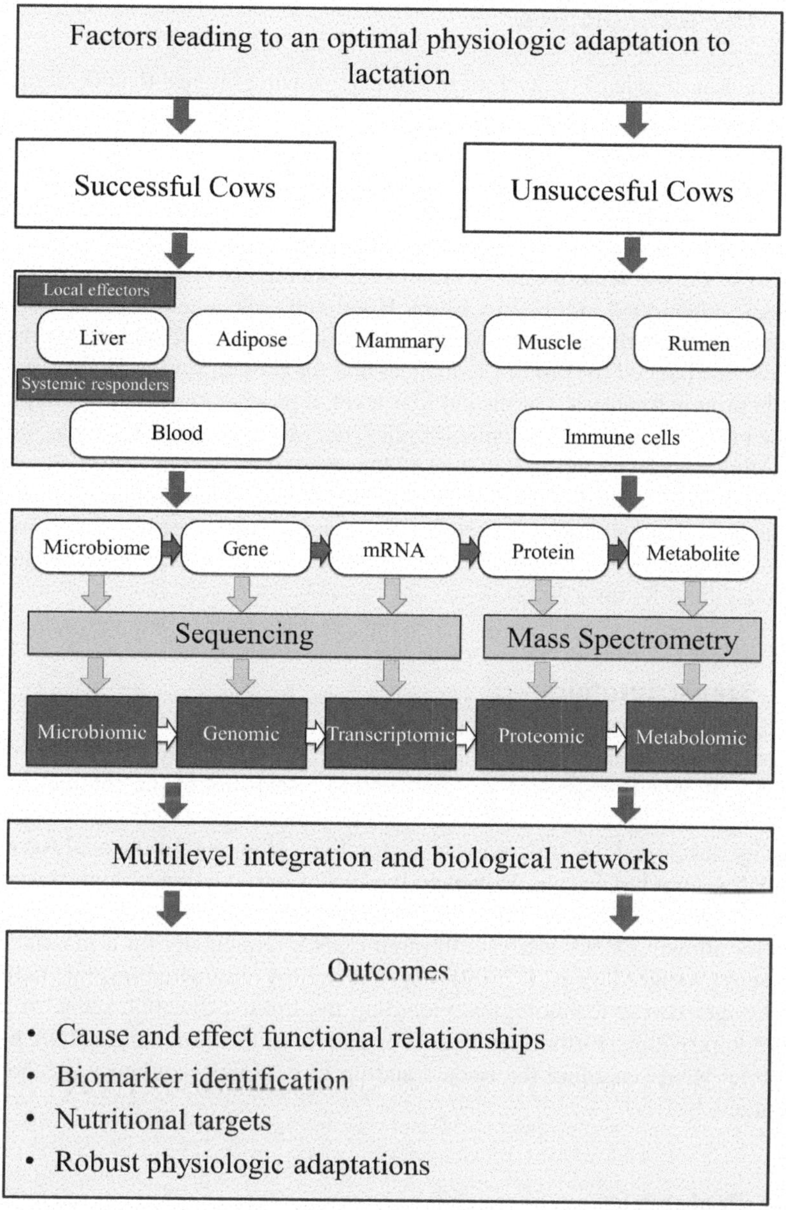

Fig. 1 Conceptual framework for the application of systems biology approaches to better understand the underlying mechanisms associated with successful and unsuccessful physiologic adaptations of lactation

2 The Technologies

During the last decade of the twentieth century, growing interest to understand how genes control physiological processes in the body motivated scientists to initiate the Human Genome Project. Due to the massive amount of sequences along the whole genome, there was a need for developing faster high-throughput screening techniques for detecting small cellular molecules, identifying hundreds of thousands of genes, and the protein and metabolite products (Loor et al. 2013b).

The first "omics" technologies were the automated DNA sequencer and the ink-jet DNA synthesizer developed by Leroy Hood and colleagues as a tool for global gene expression analysis, also known as "transcriptomics" (Hood 2002). Hood's group also introduced the protein sequencer and the protein synthesizer to study the available protein expression at the cellular level, a process known as "proteomics." With the emergence of metabolomics studies, i.e. investigating the cellular metabolite profiling, started by Frank Baganz and his group (Oliver et al. 1998), the field of "systems biology" aimed at piecing together information including transcriptomics, proteomics, and metabolomics. The goal was to understand the "big picture" within the whole system by tracking the metabolic flux from gene expression to metabolites (Winter and Kromer 2013).

2.1 Transcriptomics

The transcriptome is the total expressed RNA (i.e. mRNA, non-coding RNA, rRNA, and tRNA) in a cell or tissue, thus representing a snapshot of the cellular metabolism. The transcriptome era started when Schena et al. (1995) from Stanford University developed the "microarray" technology, allowing for the analysis of cellular mRNA on a large scale. However, the recent introduction of high-throughput next-generation DNA sequencing (NGS) technology has revolutionized transcriptomics by allowing RNA analysis through cDNA sequencing on a massive scale (RNA-seq) (Voelkerding et al. 2009). This technology eliminated several challenges posed by microarray technologies, including the limited dynamic range of detection, while providing further detailed information on the non-coding RNA portion of the total RNA, enabling the understanding of complex regulatory mechanisms (e.g. epigenetics).

2.2 Proteomics

The term "proteome" was coined by Wasinger et al. (1995) and was defined as the characterization and quantification of all sets of proteins in a cell, organ, or organism at a specific time. A proteome analysis provides the protein inventory of a cell or tissue at a defined time point, facilitating the discovery of novel biomarkers, identification and localization of posttranslational modifications, and study of protein–protein interactions (Chandramouli and Qian 2009). In the past decade, major developments in instrumentation and methodology for proteomics have been achieved (May et al. 2011). The core of modern proteomics is mass spectrometry (MS) (Aebersold and Mann 2003), a technique in which the chemical compounds in a sample are ionized and the resulting charged molecules (ions) are analyzed according to their mass-to-charge (*m/z*) ratios. One- or two-dimensional polyacrylamide gel electrophoresis (1D-PAGE, 2D-PAGE) is often used for simple preseparation of complex protein mixtures before MS analysis. To further enhance automation in the process, different types of liquid chromatography (LC or HPLC) are used to complement or substitute gel-based separation techniques.

2.3 Metabolomics

Metabolomics consists of the global profiling of metabolites utilizing high-resolution analysis together with statistical tools such as principal component analysis (PCA) and partial least squares (PLS) to derive an integrated picture of metabolites (Zhang et al. 2012). The wide spectra of molecules detectable by this approach includes peptides, amino acids, nucleic acids, carbohydrates, organic acids, vitamins, polyphenols, alkaloids, and inorganic species. Metabolome analysis may be conducted on a variety of biological fluids and tissue types and may utilize a number of different technology platforms. As one of the most common spectroscopic analytical techniques, nuclear magnetic resonance (*NMR*) can uniquely identify and simultaneously quantify a wide-range of organic compounds in the micro-molar range, providing unbiased information about metabolite profiles. Application of MS is gaining increased interest in high-throughput metabolomics, often coupled with other techniques such as chromatography (GE-MS, LC-MS, UPLS-MS) or electrophoretic techniques (CE-MS). Due to its high sensitivity and wide range of covered metabolites, MS has become the technique of choice in many metabolomics studies.

2.4 Microbiome Analysis

Although various culture-dependent approaches were traditionally used to investigate the rumen microbiota (Russell and Rychlik 2001), over 90% of the microbial communities are uncultivable due to their sensitivity to the extra-ruminal environment, even when utilizing anaerobic cell culture chambers. This limitation has for decades slowed our understanding of the complexity of the rumen microbiota. With the rise of omics-based technologies in the 1990s, culture-independent techniques to evaluate the ecology and its functional relevance such as qPCR, pyrosequencing of the 16S ribosomal RNA gene, metagenomics, and metatranscriptomics have been recently applied to address issues related to ruminant nutrition as an important means of predicting microbiota responses to dietary changes (McCann et al. 2014).

3 The Metabolic and Genetic Regulation of Intake

Voluntary feed intake is an essential parameter for all livestock enterprises, and is at the center of an animal's ability to express its full genetic potential for a productive purpose. Feed intake control and management plays a strategic role in dairy cows, especially during the transition to lactation. As stated by Drackley (1999), "the primary challenge faced by cows is a sudden and marked increase of nutrient requirements for milk production, at a time when DMI, and thus nutrient supply, lags far behind." Due to its multifactorial nature, and taking advantage of the broad spectrum of the omics technologies, DMI regulation is the perfect target for a systems approach. This said, omics technologies have not yet been widely used to investigate feed intake regulation.

The control of feed intake is a complex process that results from the integration of multiple short- and long-term signals at the feed intake regulatory centers in the brain (Morton et al. 2006). As the central nervous system is the main player in DMI regulation (Ahima and Antwi 2008), scientists started focusing on its physiology in dairy cattle. For example, comparing the prepartum cerebrospinal fluid proteome with the postpartum profile allowed Kuhla et al. (2015) to conclude that in early lactation, the pathway involved in the processing of prohormone convertase PC2 is important for the activation of various propeptides controlling feed intake and energy homeostasis. The authors particularly emphasized the importance of an increased amount of neurosecretory protein VGF, proenkephalin-A, and secretogranins, and an increased tone of endogenous opioids associated with the postparturient increase of feed intake (Kuhla et al. 2015).

Food (or energy) restriction can be used experimentally to gain insights into the mechanisms of controlling energy homeostasis and food intake regulation. Hypothalamic samples from feed restricted cows were collected and subjected to proteomic analysis (Kuhla et al. 2007). The data revealed not only lower availability of substrates but also that oxidative stress plays a role in regulating hypothalamic hormones. Synthesis of reactive oxygen species has been previously associated with reduced performance of dairy cows during the transition period (Abuelo et al. 2015). Because peripheral tissues can participate in their production, they can indirectly contribute to the regulation of feed intake. For example, not only the small and large intestine but also other gastrointestinal tract components (e.g. rumen, abomasum), liver, muscle, adipose, and other splanchnic organs express peptides such as PYY and GCG, thus, could affect the CNS and regulate intake (Pezeshki et al. 2012).

Adipose tissue has historically been a central player in the context of peripheral organs and their ability to regulate intake and energy homeostasis. In fact, gene profiling in dairy cows has revealed that the secretion of adipokines, such as leptin and resistin, is responsive to the metabolic status of the cow during the transition period or a period of negative energy balance (Friedrichs et al. 2016). Work in non-ruminants has clearly demonstrated that the effects of leptin on food intake and energy homeostasis are mediated via a network of orexigenic and anorexigenic neuropeptides (Ingvartsen and Boisclair 2001). In dairy cows, plasma resistin levels are high 1 week after calving and positively correlated with plasma NEFA levels and negatively correlated with milk yield, DMI and energy balance (Reverchon et al. 2014). Thus, the action of resistin on control of feed intake might be more connected to its metabolic and lipolytic effects (Reverchon et al. 2014).

In addition to plasma metabolites and hormones participating as humoral signals in the control of feed intake, oxidative metabolic processes in peripheral organs can also generate signals to terminate feeding. The periprandial pattern of fuel oxidation can, in fact, be involved in short-term regulating feeding behavior in the bovine (Allen et al. 2005; Allen et al. 2009). Furthermore, through the use of indirect calorimetry, and milk and plasma analysis, Derno et al. (2013) were able to demonstrate how each single feed intake event induced a nearly constant time-delayed change in net carbohydrate (COX_{net}) and net fat oxidation (FOX_{net}) in late lactation Holsteins fed ad libitum. Cows seemed to initiate feed intake in response to an accelerated FOX_{net} rate and a decelerated COX_{net} rate, respectively. As postprandial increases in COX_{net} and FOX_{net} coincide with times in which cows did not eat, the occurrence of metabolic oxidative processes was hypothesized to signal feed intake suppression, which lends support to the hepatic oxidation theory of Allen et al. (2009). The relationships among peripheral tissue and the central nervous system, and their ability to intercommunicate in the regulation of feed intake, call for the systems approach as the only suitable way to pursue the understanding of this complex topic as a way to gain further knowledge for future application in nutritional strategies.

Genetic merit analyses have recently been conducted to study feed intake regulation as a way to determine the existence of a genetic component that could, in the future, be exploited for the development of focused selection strategies (Tetens et al. 2014; Shonka et al. 2015). Spurlock et al. (2012) first demonstrated how both DMI and energy balance likely respond to selection pressure. Subsequently, Rocco and McNamara (2013) confirmed, for example, how regulation of adipose tissue metabolism in lactation is a function of both diet and genetic merit and is controlled by multiple mechanisms including gene transcription and posttranslational protein modifications. These data established the foundation for selection of animals specifically for feed intake, opening the possibility to generate genetic lines with high and low DMI, hence, allowing for direct comparisons of the underlying regulatory mechanisms controlling feed intake.

4 Immunonutrition: A Future Frontier

Perhaps because of a lack of formal training outside classical concepts in nutrition, ruminant nutritionists primarily focus on meeting the production requirements of dairy cows, without truly dissecting what components of maintenance requirement might affect performance. There is substantial evidence indicating that the immune system is intimately involved with other mechanisms that allow cows to adjust quickly to the onset of lactation without suffering chronic disorders. In fact, cows that lag behind the rest of the herd in terms of production outcomes (including fertility) often display a greater inflammatory status and compromised liver function (Bionaz et al. 2007; Bertoni et al. 2008; Trevisi et al. 2012). The lower DMI of the health-impaired animals (Trevisi et al. 2012) is not surprising, because inflammatory molecules often have anorexogenic effects (Plata-Salaman 1998, 2001; Wong and Pinkney 2004). Because it is now generally agreed that the postpartum negative energy balance is mainly caused by the reduction in DMI, rather than the increase demand of the mammary gland (Grummer et al. 2010), health and immunity has to be a focus of nutritionists aiming to improve the adaptation of the cow to lactation (LeBlanc 2010).

Several studies reported that both the innate and adaptive immune systems in peripartal cows are often compromised; for example, cytokine production is impaired (Sordillo and Babiuk 1991; Ishikawa et al. 1994), oxidative burst activity is reduced (Dosogne et al. 1999), and consequently phagocytic activity by leukocytes is often (Ingvartsen et al. 2003), but not always (Sander et al. 2011; Graugnard et al. 2012), reduced. To better understand the parturition "effect" on cow defense mechanisms, Burton et al. (2001) hybridized leucocyte RNA harvested 6 h after parturition to a custom oligo microarray and compared the gene expression to data obtained at 2 weeks prepartum. They uncovered that parturition repressed the expression of genes involved in the classic immune response as well as other genes known to be important in normal cell growth, metabolism, and responsiveness to the blood environment. Therefore, they concluded that parturition influences the expression of multiple leukocyte genes required for normal functioning of these cells, a fact that could easily explain the dysfunctional capacities of leukocytes from periparturient cows. The authors, however, did not determine which population of leukocytes was affected or what factors of parturition caused the repression of gene expression. Preliminary studies at that time implicated neutrophils as the main target of parturition-induced gene expression changes (Madsen et al. 2002; Weber et al. 2001), and currently most of the research efforts are focused on this specific population. For example, other microarray data revealed how neutrophils undergo great stress around parturition, a feature highlighted by the increased expression of antiapoptotic genes, as if these cells were trying to counteract a reduction in viability (Madsen et al. 2004). A more focused expression analysis further described how the changes in neutrophils activity observed around calving appeared related, at least in part, to alterations in purinergic signaling as well as changes in expression of genes associated with adhesion and chemoattractant binding (Seo et al. 2013).

Pharmaceutical treatment through the use of different drugs (e.g. non-steroidal anti-inflammatory compounds) have been used in different scenarios to neutralize the inflammatory state displayed, at different degrees, by almost every dairy cow, albeit with different results (Bertoni et al. 2004; Farney et al. 2013; Meier et al. 2014). Nutritional strategies, even if sometimes ineffective, such as level of dietary energy in the dry-period diet (Graugnard et al. 2012; Zhou et al. 2015), dietary amino acid balance (Yuan et al. 2014), or natural additives (Garcia et al. 2015) have resulted in some positive effects on the immune function. From a consumer standpoint, much more attentive now than before to the use of drug in food animals, nutritional approaches to boost the immune system and reduce incidence of disorders are seen in a positive light.

As the metabolic and immune networks are deeply connected (Mathis and Shoelson 2011), a systems approach clearly is beneficial when attempting to understand the underlying mechanisms elicited by different immunonutrition strategies. Because immune cell function seems to be dysfunctional during early lactation, the feed additive industry has placed some emphasis on developing "immunostimulants" as dietary supplements. For example, recent results indicated a positive role of a commercially available immunostimulant (OmiGen-AF®, Phibro Animal Health Corporation, USA) in enhancing leukocyte function which would provide added antibacterial capacity during the peripartum (Nace et al. 2014). A transcriptome analysis, coupled with the more traditional Western blot procedure, revealed how these improvements might be due to changes in expression that might alter neutrophil apoptosis, signaling, sensitivity, and response (Wang et al. 2007, 2009).

Specific dietary treatments that have been successful in improving leucocyte activity include methionine and other rumen-protected amino acids (Osorio et al. 2013a; Yuan et al. 2014), dietary nitrogen level (Raboisson et al. 2014), alpha1-acid glycoprotein (Rinaldi et al. 2008), and orange oil (Garcia et al. 2015). However, the molecular technologies (i.e. holistic approach) are mostly helping us understand some of the "mistakes" in what is considered standard in today's dairy management systems. The most resounding case regards dry cow management, and its effect, among others, on cow health status and immunity. Traditional management provides "far-off" dry cows with a high-fiber/low-energy density diet, while in the last month of gestation ("close-up" dry period) the diet increases in energy density with a lower fiber content. However, studies from different research groups have demonstrated that prepartum overfeeding of energy often results in prepartum hyperglycemia and hyperinsulinemia and marked postpartum adipose tissue mobilization (i.e. greater blood NEFA concentration) (Rukkwamsuk et al. 1999; Holtenius et al. 2003; Janovick et al. 2011; Ji et al. 2012, 2014; Khan et al. 2014). In addition, higher-energy close-up diets have also been associated with negative effects on postpartum health indices, underscoring the possible detrimental effects of this management approach (Dann et al. 2006; Soliman et al. 2007; Graugnard et al. 2013; Shahzad et al. 2014).

Transcriptome profiling of neutrophils revealed that allowing cows free access to higher-energy diets during late pregnancy resulted in the alteration of genes encompassing pathways associated with the immune response (Moyes et al. 2014; Zhou et al. 2015). Furthermore, phagocytosis activity of these cells was impaired, and early prepartal activation of inflammatory genes suggested a chronic state of compromised health (Moyes et al. 2014; Zhou et al. 2015). Transcriptomic studies further highlighted how this practice not only impairs immune function, but affects the whole system of the cow as indicated by alterations in endoplasmatic reticulum stress in hepatocytes, probably as a consequence of a higher activation of inflammatory-related functions (Shahzad et al. 2014). These omics data also revealed a predisposition of cows to fatty liver while compromising overall liver health during the periparturient period, as indicated by pro-inflammatory gene expression (Loor et al. 2006).

A surprising response of the higher-energy feeding approach during the close-up is that carry-over effects of this type of diet can be detected in the offspring of the cows (Osorio et al. 2013b), albeit with an opposite positive effect, because calves from dams fed the higher-energy prepartum had greater neutrophil phagocytosis after birth (Osorio et al. 2013b). Other studies have also indicated how applying an increased plane of nutrition directly to the neonate (and not the mother) can have beneficial effects on the calf immunity (Obeidat et al. 2013; Ballou et al. 2015). Although the occurrence of an "epigenetic" mechanism leading to these effects has not been demonstrated, Chang et al. (2015) were able to provide some evidence of how a high-concentrate diet could activate epigenetic mechanisms that contribute to the expression of immune-related genes in the livers of dairy cows. The data revealed how chromatin decompaction and DNA demethylation in relevant areas of the promoter of candidate immune genes are strongly correlated with their upregulation of expression. In light of these recent findings, further studies are needed to better understand how dietary strategies could not only affect the cow, but also the calf (i.e. the future productive animal).

The beneficial effects of rumen-protected methionine supplementation during the transition period could be used to counteract the negative outcome of a high-energy plane of nutrition during the dry period. In fact, when supplemented with rumen-protected methionine, the temporal adaptations in expression of PMN genes related with migration, development and cellular antioxidants indicated effectiveness in alleviating the negative effects of prepartal energy-overfeeding (Li et al. 2016). Furthermore, the similar DMI and milk yield of those cows compared with cows fed the lower-energy diet strengthened the idea that methionine helps overcome the limitations of overfeeding energy during the prepartal period (Li et al. 2016).

5 Systems Biology Concept in Animal Nutrition and Physiology

There is compelling evidence that integrating both omics and bioinformatics tools in peripartal cow nutrition will enhance our understanding of the complex biological functions, interactions, and adaptations among key organs (Loor et al. 2013a, b, 2015). Furthermore, the omics approach also will help unmask the complex composition of the gut microbial ecosystems and their relation to animal health and milk production. Because of its dynamic nature, application of these tools during the transition period from pregnancy through the onset lactation will result in the biggest gain in knowledge. In turn, the accumulated knowledge would enable nutritionists in the medium-to-long term to develop more effective diets tailored to enhance animal health, improve the rumen microbial profile, reduce metabolic disorders, and optimize milk production. Below are examples of published work since 2013 in which a systems approach was used to study individual tissues or the rumen microbiome (Table 1).

Table 1 Summary of published studies since the reviews of Loor et al. (2013a, b) utilizing omics tools and bioinformatics to examine the influence of nutrition on tissues or the rumen microbiome in dairy cows

Tissue	Animal model	Dietary treatment	Platform	Key biological responses	Reference
Liver	Peripartum cows	Energy restriction	cDNA microarray	Energy restriction increased lipid and amino acid catabolism both pre- and postpartum; it primed the liver to better face the early postpartum metabolic and inflammatory challenges	Shahzad et al. (2014)
	Lactating cows	Energy restriction	Oligonucleotide microarray	Hepatic gluconeogenesis, cytoskeletal remodelling, and glucose-sparing were upregulated	Grala et al. (2013)
	Early lactation cows	Energy restriction	miRNA-seq	Hepatic miRNA are associated with the expression of genes involved in lipid and energy metabolism	Fatima et al. (2014a)
	Early lactation cows	Energy restriction	miRNA array	Hepatic miRNA play a role in lipid and energy metabolism	Fatima et al. (2014b)
	Early lactation cows	Energy restriction	RNA-seq	Cytochrome p450 family 11 subfamily A member 1 (*CYP11A1*) plays a role in hepatic lipid metabolism	McCabe et al. (2012)

(continued)

Table 1 (continued)

Tissue	Animal model	Dietary treatment	Platform	Key biological responses	Reference
Rumen epithelium	Young calves	Grain-based calf starter	Oligonucleotide microarray	Identified molecular markers controlling differentiation and growth of rumen epithelium	Connor et al. (2013)
	Young calves	High-protein milk replacer (28.5% crude protein) plus high-crude protein starter (25.5% crude protein)	Oligonucleotide microarray	Weaning increased overall metabolism in rumen epithelium, especially the biosynthesis of glycan and nucleotide metabolism	Naeem et al. (2014)
	Peripartum cows	Transition from low-energy diet precalving to high-energy diet postcalving	Affymetrix microarray	Rumen papillae adapt to dietary energy changes by altering the papillae transcriptome profiles	Steele et al. (2015)
Rumen microbiota	Peripartum cows	Transition from high-forage diet precalving to high-grain diet postcalving	Pyrosequencing	High-forage diets enriched fungal communities; while high-grain diets increased the prevalence of protozoa	Kumar et al. (2015), Lima et al. (2015)
	Peripartum cows	Transition from high-forage diet precalving to high-grain diet postcalving	Terminal restriction fragment length polymorphism (TRFLP) analysis	During the transition period Streptococcaceae increased but Veillonellaceae decreased. After calving, Lactobacillaceae increased	Wang et al. (2012)

5.1 Liver

A recent study provided novel data on the hepatic transcriptome adaptations to feed restriction during early lactation and reduced milking frequency in grazing cows (Grala et al. 2013). Cows that were underfed for 3 weeks, i.e. were energy-restricted, had a marked activation of glucose-sparing pathways as well as gluconeogenesis. Genes involved in hepatic stress (e.g. angiopoietin-like 4, *ANGPTL4*; glutathione peroxidase 3, *GPX3*) were upregulated in response to the energy restriction underscoring the pressure energy restriction places on liver function (Grala et al. 2013) even in grazing cows which produce less milk than contemporary confinement-fed cows. It was noteworthy that the "cytoskeletal remodeling" pathway also was activated, indicating that enhanced tissue remodeling was a feature of hepatic adaptations to energy restriction. From a nutritional standpoint, the marked inhibition of "vitamin B6 metabolism" and "biosynthesis of unsaturated fatty acids" during feed restriction underscored the existence of key pathways that could be controlled via nutrition, e.g. dietary supplementation, to help alleviate stress on the liver during periods like the transition into lactation when all cows experience a normal decline in feed intake.

The extensive bioinformatics and gene network analyses of the hepatic transcriptome performed by Shahzad et al. (2014) revealed that feed restriction prepartum had a positive effect on the liver from the standpoint of overcoming the postpartal metabolic and inflammatory challenges. For example, the analysis of metabolic pathways revealed that energy restriction helped prime the liver by activating pathways related with the utilization of fatty acids and amino acids through ketogenesis and gluconeogenesis. Furthermore, the liver of energy-restricted cows seemed to have a higher capacity to cope with endoplasmic reticulum stress, which may lead to a decrease in hepatic protein synthesis. The hepatic inflammatory-related response also was activated by feed restriction through the upregulation of genes involved in the acute-phase response.

The upstream gene network analysis performed by Shahzad et al. (2014) uncovered that healthy liver adaptations associated with energy restriction might be controlled by stimulating lipid-related transcription factors involved in fatty acid oxidation and cell stress including the peroxisome proliferator-activated receptors (PPARs) and nuclear factor (erythroid-derived 2)-like 2 (*NFE2L2*). In contrast to these seemingly positive adaptations in the transcriptome, overfeeding during the prepartum period activated hepatic triacylglycerol synthesis and lipid accumulation, leading to mild hepatic lipidosis after parturition. The authors discussed how greater lipid infiltration could have activated cell proliferation and cell-to-cell communication pathways in the liver.

MicroRNA (miRNA) are small non-protein coding RNA molecules containing 19–22 nucleotides which act as endogenous posttranscriptional regulators of gene expression (e.g. glucose homeostasis and insulin signaling) (Dumortier et al. 2013) and play important regulatory roles in metabolism (Alisi et al. 2011). Recent studies have used RNA-seq technology to screen the hepatic miRNA expression profiles in early lactating cows undergoing negative energy balance (McCabe et al. 2012;

Fatima et al. 2014a, b). The most abundant miRNA in the liver are involved in glucose and insulin metabolism, e.g. miR-122, miR-192, miR-3596, let-7c, let-7i, let-7g, and let-7f. Furthermore, miR-17, miR-31, and miR-140 were upregulated during negative energy balance and have been associated with hepatic disorders such as fatty liver, oxidative stress, cholestasis, fibrosis, dysplasia, and cirrhosis. Clearly, application of RNA-seq provided additional information on the role of small RNA molecules in the control of transcription and liver function during negative energy balance after parturition. Additional studies would have to be performed to determine the functional outcome of the changes in miRNA profiles.

5.2 Adipose Tissue

Relative to transcriptome studies in liver, few studies have evaluated the transcriptome during the peripartum period in adipose tissue (Sumner-Thomson et al. 2011; Bionaz and Loor 2012). Data from those studies were discussed by Loor et al. (2013a, b) and will not be discussed in this chapter. Among the qPCR-based studies published recently to detect the effects of dietary energy prepartum and degree of adiposity ("body condition score," BCS) at parturition, our group revealed a negative effect of feeding a high-energy diet prepartum to over-conditioned cows (BCS 5, on a 10-point scale) through driving adipogenesis by upregulating the expression of fatty acid synthase (*FASN*), leptin, and proadipogenic miRNA such as miR-378 and miR-143 (Vailati-Riboni et al. 2016). In addition, this study revealed an upregulation of expression of chemokine ligand 5 (*CCL5*) and the cytokines interleukin 6 (*IL6*) and tumor necrosis factor (*TNF*), which indicated a pro-inflammatory response after parturition. The authors discussed the link among these changes and the increase in lipolysis, which often leads to higher susceptibility for hepatic triacylglycerol accumulation and impaired liver functions (Vailati-Riboni et al. 2016).

Saremi et al. (2013) and Haussler et al. (2015) reported little or no effect of dietary supplementation of conjugated linoleic acid (CLA) for 105 days postpartum on the expression of serum amyloid A3 (*SAA3*) (an adipokine associated with insulin resistance) and monocyte chemoattractant protein-1 (*MCP-1*) (a chemokine synthesized by adipocytes) in adipose tissue. Other PCR-based studies found no effects of feeding a high-energy diet prepartum on the expression nutrient-sensing receptors such as hydroxycarboxylic acid receptor 2 (*HCAR2*) (Friedrichs et al. 2016) or the Sirtuin-1 (*SIRT1*) PPARγ co-activator 1α (*PPARGC1A*) axis in adipose tissues (Weber et al. 2016).

5.3 Immune Cells

Sasaki et al. (2014) used microarray technology to identify the affected gene networks in peripheral blood mononuclear cells associated with hypocalcemia in dairy cows. The authors uncovered 98 affected genes in cows afflicted with hypocalcemia and among those the expression of protein kinase (cAMP-dependent, catalytic) inhibitor β (*PKIB*), DNA-damage-inducible transcript 4 (*DDIT4*), period homolog 1 (*PER1*), and NUAK family, SNF1-like kinase 1 (*NUAK1*) were closely associated with both experimental hypocalcemia and milk fever. Although the authors did not perform a bioinformatics analysis of the data, the results support the view that "the effect of hypocalcemia on the mRNA expression of these genes in the tissues that regulate calcium homeostasis in dairy cows should be determined."

5.4 Rumen Epithelium

Recent transcriptome studies revealed that preweaning development is associated with activation of pathways related to cell morphology, cell death, cell cycle, cellular growth, cellular proliferation, molecular transport, and lipid metabolism, while the postweaning stage is more associated with activation of cell adhesion molecules, p53 signaling, and fatty acid metabolism pathways (Connor et al. 2013; Naeem et al. 2014). Early nutrition in dairy calves also could play a vital role in modulating gene expression in rumen papillae. For example, introducing high-protein milk replacer (containing 28.5% crude protein) plus high-crude protein starter (containing 25.5% crude protein) to dairy calves until 42 days of age improved rumen development and elicited more efficient utilization of nutrients (Naeem et al. 2014). The latter was surmised through the bioinformatics analysis of transcriptome data revealing the activation of "carbohydrate metabolism" through inducing citrate cycle, galactose metabolism, butanoate metabolism, glycolysis/gluconeogenesis, and pyruvate metabolism (Naeem et al. 2014). Furthermore, lipid metabolism also was activated primarily because of the induction of steroid biosynthesis, sphingolipid metabolism, and biosynthesis of unsaturated fatty acids (Naeem et al. 2014). In another study dealing with the transcriptome adaptations of rumen epithelium to feeding grain- versus hay-based diet, Connor et al. (2014) uncovered that feeding grain enhanced the development of the rumen papillae through activating transcription factors such as transforming growth factor β1 (*TGFB1*), forkhead box O1 (*FOXO1*), and peroxisome proliferator-activated receptor alpha (*PPARA*) transcription factors. In contrast, feeding hay provided more available energy to rumen epithelium in the form of butyrate production. As a result it was hypothesized that butyrate could have stimulated the expression of the transcription factor estrogen-related receptor alpha (*ESRRA*), hence, playing a role in energy metabolism of the developing rumen (Connor et al. 2014).

Transcriptome analysis using microarrays revealed that rumen epithelium adapts to a high-energy diet after parturition by activating epidermal growth factor signaling (*EGFR*), growth hormone receptor (*GHR*), and transforming growth factor β1 (*TGFB1*) pathways, all of which may enhance VFA utilization by rumen epithelium (Steele et al. 2015). Additional genes that appear to have a role in the adaptations of the rumen epithelium to lactation include an upregulation of 3-hydroxy-3-methylglutaryl-CoA synthase 2 (*HMGCS2*) along with the downregulation of the transcription factor retinoid X receptor α (*RXRA*), the insulin receptor (*INSR*), and ribosomal protein S6 kinase (*RPS6KB1*) (Minuti et al. 2015). Because of its sensitivity to changes in dietary management, it is expected that transcriptome analysis of the rumen epithelium during the transition into lactation will continue in the future. More importantly, it will be important to address more explicitly the link between diet, the rumen microbiome, and the rumen epithelium.

5.5 Ruminal Microbes

In commercial dairy farms, high-producing cows are shifted from low-energy diet before calving to high-energy diet postcalving to provide the rumen microbial communities with more readily available energy for enhancing the microbial fermentation in the rumen. As a result, rumen microbes generate more energy in the form of VFA which is required for milk synthesis (Roche et al. 2013a). In the first study of its kind, Wang et al. (2012) using terminal restriction fragment length polymorphism (TRFLP) analysis of the 16S rRNA gene in whole rumen digesta DNA revealed that members of the Streptococcaceae family increased while Veillonellaceae decreased during the transition period compared with 100 days postcalving. Furthermore, members of the Lactobacillaceae family were more abundant after calving. More recent studies employing pyrosequencing technology revealed that changes in dietary energy alter the composition, structure, and diversity of the rumen microbiota. For example, Kumar et al. (2015) and Lima et al. (2015) reported that enrichment of fungal communities were associated with higher dietary fiber in the prepartum, while the prevalence of protozoa, associated with starch digestion, increased with higher dietary energy in the postpartum.

6 The Physiologic Implication of Body Fatness (BCS)

In dairy management systems, BCS is used as an indicator of body fat content and cow nutritional status. Cows should be managed to achieve appropriate BCS both pre- and postpartum to reduce threats to welfare, because BCS at calving may affect early lactation DMI, postcalving BCS loss, milk yield, cow immunity, and fertility (Roche et al. 2009). At calving, DMI and BCS are negatively correlated (Hayirli et al. 2002; Matthews et al. 2012), so that "fat" cows undergo a more pronounced and prolonged depression in DMI, leading to a deeper negative energy balance (Hayirli et al. 2002; Agenas et al. 2003), an increase in lipomobilization, and a greater and persistent increase in blood NEFA (Dann et al. 2006). However, not only over-conditioned but also under-conditioned dairy cows have a greater incidence of diseases than animals with a normal BCS (Roche et al. 2009). The reproductive side is also influenced, as low BCS is a risk factor for postpartum endometritis and delayed cyclicity in dairy cows (Kadivar et al. 2014). Gene expression analysis further investigated the low reproduction results in obese heifers, suggesting how, unlike normal and lean cows, obese cows had suppressed granulocyte macrophage colony-stimulating factor gene expression (an embryogenesis promotant) in the ampulla (Nahar et al. 2013). Furthermore, the cow BCS has been shown to influence calving itself, with higher BCS increasing chances of dystocia, and to correlate with calf survival (Bastin et al. 2010).

Although these phenotypic effects of calving BCS are known, not much has been done to understand the molecular mechanisms behind them. As for DMI, geneticists determined that selection for good body condition, body conformation, and optimal milk production is possible and their genetic associations could be useful for designing breeding goals (Koenen et al. 2001; Kadarmideen and Wegmann 2003). This underscores that the regular measurement of first-lactation BCS records should be sufficient for genetic evaluation (Loker et al. 2011). The fact that a genetic basis exists opens the possibility to evaluate the BCS effect not only with traditional metabolic parameters but also with more sophisticated techniques. Because insulin resistance is related to BCS (Kawashima et al. 2016), and an increase in prepartum adiposity can predispose dairy cows to a greater magnitude of insulin resistance during early lactation (Holtenius et al. 2003; Holtenius and Holtenius 2007), Rico et al. (2015) focused on the study of ceramide sphingolipids that are believed to mediate the inhibitory effect of saturated fatty acids on insulin signaling (Chavez et al. 2014). Metabolomics data, using MS, revealed a remodeled plasma sphingolipidome in dairy cows transitioning from late pregnancy to lactation, characterized by an accumulation of ceramides in plasma during the progression of insulin resistance in overweight cows transitioning from late pregnancy to early lactation. These data supported the potential involvement of ceramides in the pathological development of insulin resistance in dairy cattle.

Due to lack of metabolic data, an extensive amount of work has been conducted to evaluate effect of BCS in grazing systems, which normally depends on data gathered from more intensive and controlled systems (e.g. TMR-fed systems). Compared with a TMR-based system where the risk of over-conditioning during the dry period is far greater, in a pasture system, thin cows have a greater prevalence of problems (Roche et al. 2007, 2009). For these reasons, assessment of BCS prepartum can provide a qualitative evaluation of the chances for an optimal transition, which, in turn, is closely associated with optimal production and the chances for successful lactation (Roche et al. 2009).

Roche et al. (2013c) were the first to provide evidence linking BCS to cow health and welfare. Low peripartum BCS was, in fact, associated with alterations in albumin, urea, and magnesium metabolism, which altogether may place cows at a greater risk of developing subclinical endometritis. Further, biomarkers of liver function and the acute phase response indicated that BCS did not affect the cows' ability to mount inflammatory responses to the stimuli encountered in the peripartum and during early lactation. Based on these results, the authors concluded that BCS alone is not a sufficiently sensitive measure to be reflective of cow welfare. Regardless, these results supported the general recommendations that a calving BCS of 4.5 to 5.5 (10-point scale) would optimize production, reproductive performance, and general health.

To further understand the effect of BCS on the whole organism, transcriptomic analysis was performed on the liver of these animals. The microarray output revealed how calving BCS mostly induces changes in metabolic pathways. Thinner cows had a compromised gluconeogenic capacity, with increased glucose use, that could probably explain the lower production performance (Akbar et al. 2014). However, both thin and over-conditioned cows displayed indices of an inflammatory state at the level of the liver (Akbar et al. 2014). This was further concluded by a more focused gene expression experiment (Akbar et al. 2015) which revealed that cows calving at BCS 3.5 or 5.5 were compromised relative to cows calving at BCS 4.5 (10-point scale). Collectively, the results supported an optimum calving BCS for pasture-based dairy cows of approximately BCS 5.0 (i.e. equivalent to BCS 3.0 in a 5-point scale), similar to that recommended by Roche et al. (2009).

Despite the fact that BCS plays an important role in the metabolic response of the animal to lactation and its level is regulated through nutrition, in grazing systems cows with a different level of adiposity are generally managed similarly during the prepartum period (Roche et al. 2013a). Although several studies attempted to understand the metabolic and molecular changes associated with dietary energy intake before calving, the contributing factors remain poorly understood. The calving BCS has been proposed as a key contributing factor. To address this lack of knowledge, gene and microRNA (miRNA) expression profiling were used to better understand the interaction between precalving BCS and plane of nutrition in the metabolism of adipose tissue (Vailati Riboni et al. 2016). Specifically, how these factors influence the adipose response to the physiological changes induced by the high metabolic demands of early lactation.

Overfeeding optimally conditioned cows during the last 3 weeks before parturition primed adipose tissue for accretion of lipid and a robust localized inflammatory response, which upon parturition could increase the probability of metabolic disorders; that is, localized inflammation renders the adipocyte more susceptible to lipolytic signals that could result in greater flow of fatty acids into the liver (Vailati Riboni et al. 2016). Similarly, prepartum nutrient restriction of thinner cows enhanced the localized pro-inflammatory response of adipocytes, hence, eliciting a similar negative outcome. Using a microarray platform, the same interaction was studied in the hepatic adaptation of these animals (unpublished data). As hypothesized, the effect of prepartum plane of nutrition on hepatic function was dependent on the BCS of the cow, underscoring how these management tools need to be evaluated together to optimize the biological adaptations of the cow during the peripartum period.

The more pronounced transcriptome changes in under-conditioned cows highlighted that they were more susceptible to prepartum feeding level/allocation than optimally conditioned cows. Similarly, the bioinformatics analysis revealed transcriptome signatures that indicated a greater and potentially more prolonged negative energy balance in overfed optimally conditioned cows and also in feed-restricted under-conditioned cows. Overall, the combined data indicated that a regimen of nutrient restriction prepartum in optimally conditioned cows avoids detrimental effects both at the adipose tissue and liver level; hence, physiologically priming the cow to the demands of lactation and avoiding a metabolically "lazy" phenotype. Instead, thinner animals seem to benefit from a higher plane of nutrition.

As these data cumulatively indicate, BCS alone as well as its interaction with feeding management have an effect on the health status of the grazing cow. In TMR-based systems, over-conditioned cows are at higher risk of infection due to the fact that the intense lipomobilization taking place around calving is associated with alterations of lymphocyte function (Lacetera et al. 2005); hence, directly linking calving BCS with the immune system. In light of this, the effect of precalving BCS and level of feeding on immunocompetence during the peripartum period has been investigated (Lange et al. 2016). Gene expression analysis of in vitro-stimulated cells revealed how cows with high energy intake precalving had increased cytokine expression precalving. Furthermore, it confirmed that optimally conditioned animals might benefit from a restricted diet precalving, whereas under-conditioned cows could be fed to requirements.

A systems approach, encompassing hepatocyte and adipocyte metabolism, and immune cell responses, has been a valuable tool for determining calving BCS recommendations for grazing cows. Furthermore, because these recommendations could not always be met, the systems approach was found very reliable in the evaluation of how different levels of precalving (calving) adiposity can interact with different nutrient supply to better define nutritional management of the dry grazing cow.

7 Perspectives

A vast amount of knowledge has been acquired since the review of Drackley (1999), outlining crucial gaps in knowledge on transition dairy cow biology. Omics technologies have contributed widely to the understanding of the delicate physiologic equilibrium that allows for a successful transition into lactation. However, we still tend to consider single organs as the "system" to study, subsequently inferring the connection with the rest of the organism based on the existing literature. The systems approach in its most pure connotation has not yet been applied to the study of dairy cows in the context of nutritional management and its role in the animal's adaptations to lactation. This is largely due to the fact that when integrating multiple datasets, one tends to generate bare numerical relationships rather than meaningful biological connections among organs. Therefore, as a future frontier, the dairy science community must address the need for "useful" approaches (e.g. modelling, bioinformatics) to integrate knowledge derived from the main drivers of the adaptations to lactation, e.g. rumen microbiome and epithelium, liver, adipose, and mammary gland. It is believed that greater focus on integration among various components of the cow system will generate more meaningful biological networks to push forward knowledge on the biology of the transition cow.

References

Abuelo A, Hernandez J, Benedito JL, Castillo C (2015) The importance of the oxidative status of dairy cattle in the periparturient period: revisiting antioxidant supplementation. J Anim Physiol Anim Nutr (Berl) 99(6):1003–1016

Aebersold R, Mann M (2003) Mass spectrometry-based proteomics. Nature 422(6928):198–207

Agenas S, Burstedt E, Holtenius K (2003) Effects of feeding intensity during the dry period. 1. Feed intake, body weight, and milk production. J Dairy Sci 86(3):870–882
Ahima RS, Antwi DA (2008) Brain regulation of appetite and satiety. Endocrinol Metab Clin North Am 37(4):811–823
Akbar H, Zhou Z, Macdonald K, Schutz KE, Verkerk G, Webster JR, Rodriguez-Zas SL, Roche JR, Loor JJ (2014) Differences in hepatic transcriptional regulatory networks due to body condition score at calving in grazing dairy cattle. J Dairy Sci (97 E-Suppl. 1)
Akbar H, Grala TM, Vailati Riboni M, Cardoso FC, Verkerk G, McGowan J, Macdonald K, Webster J, Schutz K, Meier S, Matthews L, Roche JR, Loor JJ (2015) Body condition score at calving affects systemic and hepatic transcriptome indicators of inflammation and nutrient metabolism in grazing dairy cows. J Dairy Sci 98(2):1019–1032
Alisi A, Da Sacco L, Bruscalupi G, Piemonte F, Panera N, De Vito R, Leoni S, Bottazzo GF, Masotti A, Nobili V (2011) Mirnome analysis reveals novel molecular determinants in the pathogenesis of diet-induced nonalcoholic fatty liver disease. Lab Invest J Tech Meth Pathol 91(2):283–293
Allen MS, Piantoni P (2013) Metabolic control of feed intake: implications for metabolic disease of fresh cows. Vet Clin North Am Food Anim Pract 29(2):279–297
Allen MS, Bradford BJ, Harvatine KJ (2005) The cow as a model to study food intake regulation. Annu Rev Nutr 25:523–547
Allen MS, Bradford BJ, Oba M (2009) Board invited review: the hepatic oxidation theory of the control of feed intake and its application to ruminants. J Anim Sci 87(10):3317–3334
Aschenbach JR, Kristensen NB, Donkin SS, Hammon HM, Penner GB (2010) Gluconeogenesis in dairy cows: the secret of making sweet milk from sour dough. IUBMB Life 62(12):869–877
Ballou MA, Hanson DL, Cobb CJ, Obeidat BS, Sellers MD, Pepper-Yowell AR, Carroll JA, Earleywine TJ, Lawhon SD (2015) Plane of nutrition influences the performance, innate leukocyte responses, and resistance to an oral Salmonella enterica serotype Typhimurium challenge in Jersey calves. J Dairy Sci 98(3):1972–1982
Bastin C, Loker S, Gengler N, Sewalem A, Miglior F (2010) Short communication: genetic relationship between calving traits and body condition score before and after calving in Canadian Ayrshire second-parity cows. J Dairy Sci 93(9):4398–4403
Bertoni G, Trevisi E (2013) Use of the liver activity index and other metabolic variables in the assessment of metabolic health in dairy herds. Vet Clin North Am Food Anim Pract 29(2):413–431
Bertoni G, Trevisi E, Piccioli-Cappelli F (2004) Effects of acetyl-salicylate used in post-calving of dairy cows. Vet Res Commun 28(Suppl 1):217–219
Bertoni G, Trevisi E, Han X, Bionaz M (2008) Effects of inflammatory conditions on liver activity in puerperium period and consequences for performance in dairy cows. J Dairy Sci 91(9):3300–3310
Bionaz M, Trevisi E, Calamari L, Librandi F, Ferrari A, Bertoni G (2007) Plasma paraoxonase, health, inflammatory conditions, and liver function in transition dairy cows. J Dairy Sci 90(4):1740–1750
Bionaz M, and Loor JJ (2012) Ruminant metabolic systems biology: reconstruction and integration of transcriptome dynamics underlying functional responses of tissues to nutrition and physiological state. Gene Regul Syst Bio 6:109–125
Bradford BJ, Yuan K, Farney JK, Mamedova LK, Carpenter AJ (2015) Invited review: Inflammation during the transition to lactation: New adventures with an old flame. J Dairy Sci 98(10):6631–6650
Burton JL, Madsen SA, Yao J, Sipkovsky SS, Coussens PM (2001) An immunogenomics approach to understanding periparturient immunosuppression and mastitis susceptibility in dairy cows. Acta Vet Scand 42(3):407–424
Chandramouli K, Qian PY (2009) Proteomics: challenges, techniques and possibilities to overcome biological sample complexity. Hum Genomics Proteomics 2009
Chang G, Zhang K, Xu T, Jin D, Guo J, Zhuang S, Shen X (2015) Epigenetic mechanisms contribute to the expression of immune related genes in the livers of dairy cows fed a high concentrate diet. PLoS One 10(4), e0123942

Chavez JA, Siddique MM, Wang ST, Ching J, Shayman JA, Summers SA (2014) Ceramides and glucosylceramides are independent antagonists of insulin signaling. J Biol Chem 289(2):723–734

Collier RJ, McNamara JP, Wallace CR, Dehoff MH (1984) A review of endocrine regulation of metabolism during lactation. J Anim Sci 59(2):498–510

Connor EE, Baldwin RLT, Li CJ, Li RW, Chung H (2013) Gene expression in bovine rumen epithelium during weaning identifies molecular regulators of rumen development and growth. Funct Integrat Genom 13(1):133–142

Connor EE, Baldwin RLT, Walker MP, Ellis SE, Li C, Kahl S, Chung H, Li RW (2014). Transcriptional regulators transforming growth factor-beta1 and estrogen-related receptor-alpha identified as putative mediators of calf rumen epithelial tissue development and function during weaning. J Dairy Sci 97(7):4193–4207

Dann HM, Litherland NB, Underwood JP, Bionaz M, D'Angelo A, McFadden JW, Drackley JK (2006) Diets during far-off and close-up dry periods affect periparturient metabolism and lactation in multiparous cows. J Dairy Sci 89(9):3563–3577

Derno M, Nurnberg G, Schon P, Schwarm A, Rontgen M, Hammon HM, Metges CC, Bruckmaier RM, Kuhla B (2013) Short-term feed intake is regulated by macronutrient oxidation in lactating Holstein cows. J Dairy Sci 96(2):971–980

Dosogne H, Burvenich C, Freeman AE, Kehrli ME Jr, Detilleux JC, Sulon J, Beckers JF, Hoeben D (1999) Pregnancy-associated glycoprotein and decreased polymorphonuclear leukocyte function in early post-partum dairy cows. Vet Immunol Immunopathol 67(1):47–54

Drackley JK (1999) ADSA Foundation Scholar Award. Biology of dairy cows during the transition period: the final frontier? J Dairy Sci 82(11):2259–2273

Drackley JK, Donkin SS, Reynolds CK (2006) Major advances in fundamental dairy cattle nutrition. J Dairy Sci 89(4):1324–1336

Dumortier O, Hinault C, Van Obberghen E (2013) MicroRNAs and metabolism crosstalk in energy homeostasis. Cell Metab 18(3):312–324

Farney JK, Mamedova LK, Coetzee JF, KuKanich B, Sordillo LM, Stoakes SK, Minton JE, Hollis LC, Bradford BJ (2013) Anti-inflammatory salicylate treatment alters the metabolic adaptations to lactation in dairy cattle. Am J Physiol Regul Integr Comp Physiol 305(2): R110–R117

Fatima A, Lynn DJ, O'Boyle P, Seoighe C, Morris D (2014a) The miRNAome of the postpartum dairy cow liver in negative energy balance. BMC Genomics 15:279

Fatima A, Waters S, O'Boyle P, Seoighe C, Morris DG (2014b) Alterations in hepatic miRNA expression during negative energy balance in postpartum dairy cattle. BMC Genomics 15:28

Friedrichs P, Sauerwein H, Huber K, Locher LF, Rehage J, Meyer U, Danicke S, Kuhla B, Mielenz M (2016) Expression of metabolic sensing receptors in adipose tissues of periparturient dairy cows with differing extent of negative energy balance. Animal 10(4):623–632

Garcia M, Elsasser TH, Biswas D, Moyes KM (2015) The effect of citrus-derived oil on bovine blood neutrophil function and gene expression in vitro. J Dairy Sci 98(2):918–926

Grala TM, Kay JK, Phyn CV, Bionaz M, Walker CG, Rius AG, Snell RG, Roche JR (2013) Reducing milking frequency during nutrient restriction has no effect on the hepatic transcriptome of lactating dairy cattle. Physiol Genomics 45(23):1157–1167

Graugnard DE, Bionaz M, Trevisi E, Moyes KM, Salak-Johnson JL, Wallace RL, Drackley JK, Bertoni G, Loor JJ (2012) Blood immunometabolic indices and polymorphonuclear neutrophil function in peripartum dairy cows are altered by level of dietary energy prepartum. J Dairy Sci 95(4):1749–1758

Graugnard DE, Moyes KM, Trevisi E, Khan MJ, Keisler D, Drackley JK, Bertoni G, Loor JJ (2013) Liver lipid content and inflammometabolic indices in peripartal dairy cows are altered in response to prepartal energy intake and postpartal intramammary inflammatory challenge. J Dairy Sci 96(2):918–935

Grummer RR, Wiltbank MC, Fricke PM, Watters RD, Silva-Del-Rio N (2010) Management of dry and transition cows to improve energy balance and reproduction. J Reprod Dev 56(Suppl):S22–S28

Haussler S, Sacre C, Friedauer K, Danicke S, Sauerwein H (2015) Short communication: localization and expression of monocyte chemoattractant protein-1 in different subcutaneous and visceral adipose tissues of early-lactating dairy cows. J Dairy Sci 98(9):6278–6283

Hayirli A, Grummer RR, Nordheim EV, Crump PM (2002) Animal and dietary factors affecting feed intake during the prefresh transition period in Holsteins. J Dairy Sci 85(12):3430–3443

Holtenius P, Holtenius K (2007) A model to estimate insulin sensitivity in dairy cows. Acta Vet Scand 49:29

Holtenius K, Agenas S, Delavaud C, Chilliard Y (2003) Effects of feeding intensity during the dry period. 2. Metabolic and hormonal responses. J Dairy Sci 86(3):883–891

Hood L (2002) A personal view of molecular technology and how it has changed biology. J Proteome Res 1(5):399–409

Ingvartsen KL, Andersen JB (2000) Integration of metabolism and intake regulation: a review focusing on periparturient animals. J Dairy Sci 83(7):1573–1597

Ingvartsen KL, Boisclair YR (2001) Leptin and the regulation of food intake, energy homeostasis and immunity with special focus on periparturient ruminants. Domest Anim Endocrinol 21(4):215–250

Ingvartsen KL, Moyes K (2013) Nutrition, immune function and health of dairy cattle. Animal 7(Suppl 1):112–122

Ingvartsen KL, Dewhurst RJ, Friggens NC (2003) On the relationship between lactational performance and health: is it yield or metabolic imbalance that cause production diseases in dairy cattle? A position paper. Livestock Prod Sci 83(2–3):277–308

Ishikawa H, Shirahata T, Hasegawa K (1994) Interferon-gamma production of mitogen stimulated peripheral lymphocytes in perinatal cows. J Vet Med Sci 56(4):735–738

Janovick NA, Boisclair YR, Drackley JK (2011) Prepartum dietary energy intake affects metabolism and health during the periparturient period in primiparous and multiparous Holstein cows. J Dairy Sci 94(3):1385–1400

Ji P, Osorio JS, Drackley JK, Loor JJ (2012) Overfeeding a moderate energy diet prepartum does not impair bovine subcutaneous adipose tissue insulin signal transduction and induces marked changes in peripartal gene network expression. J Dairy Sci 95(8):4333–4351

Ji P, Drackley JK, Khan MJ, Loor JJ (2014) Overfeeding energy upregulates peroxisome proliferator-activated receptor (PPAR)gamma-controlled adipogenic and lipolytic gene networks but does not affect proinflammatory markers in visceral and subcutaneous adipose depots of Holstein cows. J Dairy Sci 97(6):3431–3440

Kadarmideen HN, Wegmann S (2003) Genetic parameters for body condition score and its relationship with type and production traits in Swiss Holsteins. J Dairy Sci 86(11):3685–3693

Kadivar A, Ahmadi MR, Vatankhah M (2014) Associations of prepartum body condition score with occurrence of clinical endometritis and resumption of postpartum ovarian activity in dairy cattle. Trop Anim Health Prod 46(1):121–126

Kawashima C, Munakata M, Shimizu T, Miyamoto A, Kida K, Matsui M (2016) Relationship between the degree of insulin resistance during late gestation and postpartum performance in dairy cows and factors that affect growth and metabolic status of their calves. J Vet Med Sci 78(5):739–745

Khan MJ, Jacometo CB, Graugnard DE, Correa MN, Schmitt E, Cardoso F, Loor JJ (2014) Overfeeding dairy cattle during late-pregnancy alters hepatic PPARalpha-regulated pathways including hepatokines: impact on metabolism and peripheral insulin sensitivity. Gene Regul Syst Bio 8:97–111

Koenen EP, Veerkamp RF, Dobbelaar P, De Jong G (2001) Genetic analysis of body condition score of lactating Dutch Holstein and Red-and-White heifers. J Dairy Sci 84(5):1265–1270

Kuhla B, Kuhla S, Rudolph PE, Albrecht D, Metges CC (2007) Proteomics analysis of hypothalamic response to energy restriction in dairy cows. Proteomics 7(19):3602–3617

Kuhla B, Laeger T, Husi H, Mullen W (2015) Cerebrospinal fluid prohormone processing and neuropeptides stimulating feed intake of dairy cows during early lactation. J Proteome Res 14(2):823–828

Kumar S, Indugu N, Vecchiarelli B, Pitta DW (2015) Associative patterns among anaerobic fungi, methanogenic archaea, and bacterial communities in response to changes in diet and age in the rumen of dairy cows. Front Microbiol 6:781

Lacetera N, Scalia D, Bernabucci U, Ronchi B, Pirazzi D, Nardone A (2005) Lymphocyte functions in overconditioned cows around parturition. J Dairy Sci 88(6):2010–2016

Lange J, McCarthy A, Kay J, Meier S, Walker C, Crookenden MA, Mitchell MD, Loor JJ, Roche JR, Heiser A (2016) Prepartum feeding level and body condition score affect immunological performance in grazing dairy cows during the transition period. J Dairy Sci 99(3):2329–2338

Laporta J, Moore SA, Weaver SR, Cronick CM, Olsen M, Prichard AP, Schnell BP, Crenshaw TD, Penagaricano F, Bruckmaier RM, Hernandez LL (2015) Increasing serotonin concentrations alter calcium and energy metabolism in dairy cows. J Endocrinol 226(1):43–55

Larsen M, Kristensen NB (2013) Precursors for liver gluconeogenesis in periparturient dairy cows. Animal 7(10):1640–1650

Larsen M, Galindo C, Ouellet DR, Maxin G, Kristensen NB, Lapierre H (2015) Abomasal amino acid infusion in postpartum dairy cows: effect on whole-body, splanchnic, and mammary amino acid metabolism. J Dairy Sci 98(11):7944–7961

LeBlanc S (2010) Monitoring metabolic health of dairy cattle in the transition period. J Reprod Dev 56(Suppl):S29–S35

Li C, Batistel F, Osorio JS, Drackley JK, Luchini D, Loor JJ (2016) Peripartal rumen-protected methionine supplementation to higher energy diets elicits positive effects on blood neutrophil gene networks, performance and liver lipid content in dairy cows. J Anim Sci Biotechnol 7:18

Lima FS, Oikonomou G, Lima SF, Bicalho ML, Ganda EK, Filho JC, Lorenzo G, Trojacanec P, Bicalhoa RC (2015) Prepartum and postpartum rumen fluid microbiomes: characterization and correlation with production traits in dairy cows. Appl Environ Microbiol 81(4):1327–1337

Loker S, Bastin C, Miglior F, Sewalem A, Schaeffer LR, Jamrozik J, Osborne V (2011) Short communication: estimates of genetic parameters of body condition score in the first 3 lactations using a random regression animal model. J Dairy Sci 94(7):3693–3699

Loor JJ, Dann HM, Guretzky NA, Everts RE, Oliveira R, Green CA, Litherland NB, Rodriguez-Zas SL, Lewin HA, Drackley JK (2006) Plane of nutrition prepartum alters hepatic gene expression and function in dairy cows as assessed by longitudinal transcript and metabolic profiling. Physiol Genomics 27(1):29–41

Loor JJ, Bertoni G, Hosseini A, Roche JR, Trevisi E (2013a) Functional welfare – using biochemical and molecular technologies to understand better the welfare state of peripartal dairy cattle. Anim Prod Sci 53:931–953

Loor JJ, Bionaz M, Drackley JK (2013b) Systems physiology in dairy cattle: nutritional genomics and beyond. Annu Rev Anim Biosci 1:365–392

Loor JJ, Vailati-Riboni M, McCann JC, Zhou Z, Bionaz M (2015) Triennial Lactation Symposium: Nutrigenomics in livestock: systems biology meets nutrition. J Anim Sci 93(12):5554–5574

Madsen SA, Chang LC, Hickey MC, Rosa GJ, Coussens PM, Burton JL (2004) Microarray analysis of gene expression in blood neutrophils of parturient cows. Physiol Genomics 16(2):212–221

Madsen SA, et al (2002) Altered expression of cellular genes in neutrophils of periparturient dairy cows. Vet Immunol Immunopathol 86(3–4):159–175

Mathis D, Shoelson SE (2011) Immunometabolism: an emerging frontier. Nat Rev Immunol 11(2):81

Matthews LR, Cameron C, Sheahan AJ, Kolver ES, Roche JR (2012) Associations among dairy cow body condition and welfare-associated behavioral traits. J Dairy Sci 95(5):2595–2601

May C, Brosseron F, Chartowski P, Schumbrutzki C, Schoenebeck B, Marcus K (2011) Instruments and methods in proteomics. Methods Mol Biol 696:3–26

McCabe M, Waters S, Morris D, Kenny D, Lynn D, Creevey C (2012) RNA-seq analysis of differential gene expression in liver from lactating dairy cows divergent in negative energy balance. BMC Genomics 13:193

McCann JC, Wickersham TA, Loor JJ (2014) High-throughput methods redefine the Rumen microbiome and its relationship with nutrition and metabolism. Bioinform Biol Insights 8:109–125

McNamara JP (1991) Regulation of adipose tissue metabolism in support of lactation. J Dairy Sci 74(2):706–719

McNamara JP (1997) Adipose tissue metabolism during lactation: where do we go from here? Proc Nutr Soc 56(1A):149–167

Meier S, Priest NV, Burke CR, Kay JK, McDougall S, Mitchell MD, Walker CG, Heiser A, Loor JJ, Roche JR (2014) Treatment with a nonsteroidal antiinflammatory drug after calving did not improve milk production, health, or reproduction parameters in pasture-grazed dairy cows. J Dairy Sci 97(5):2932–2943

Minuti A, Palladino A, Khan MJ, Alqarni S, Agrawal A, Piccioli-Capelli F, Hidalgo F, Cardoso FC, Trevisi E, Loor JJ (2015) Abundance of ruminal bacteria, epithelial gene expression, and systemic biomarkers of metabolism and inflammation are altered during the peripartal period in dairy cows. J Dairy Sci 98(12):8940–8951

Morton GJ, Cummings DE, Baskin DG, Barsh GS, Schwartz MW (2006) Central nervous system control of food intake and body weight. Nature 443(7109):289–295

Moyes KM (2015) Triennial Lactation Symposium: Nutrient partitioning during intramammary inflammation: a key to severity of mastitis and risk of subsequent diseases? J Anim Sci 93(12):5586–5593

Moyes KM, Graugnard DE, Khan MJ, Mukesh M, Loor JJ (2014) Postpartal immunometabolic gene network expression and function in blood neutrophils are altered in response to prepartal energy intake and postpartal intramammary inflammatory challenge. J Dairy Sci 97(4):2165–2177

Nace EL, Nickerson SC, Kautz FM, Breidling S, Wochele D, Ely LO, Hurley DJ (2014) Modulation of innate immune function and phenotype in bred dairy heifers during the periparturient period induced by feeding an immunostimulant for 60 days prior to delivery. Vet Immunol Immunopathol 161(3-4):240–250

Naeem A, Drackley JK, Lanier JS, Everts RE, Rodriguez-Zas SL, Loor JJ (2014) Ruminal epithelium transcriptome dynamics in response to plane of nutrition and age in young Holstein calves. Funct Integr Genomics 14(1):261–273

Nahar A, Maki S, Kadokawa H (2013) Suppressed expression of granulocyte macrophage colony-stimulating factor in oviduct ampullae of obese cows. Anim Reprod Sci 139(1-4):1–8

Obeidat BS, Cobb CJ, Sellers MD, Pepper-Yowell AR, Earleywine TJ, Ballou MA (2013) Plane of nutrition during the preweaning period but not the grower phase influences the neutrophil activity of Holstein calves. J Dairy Sci 96(11):7155–7166

Oliver SG, Winson MK, Kell DB, Baganz F (1998) Systematic functional analysis of the yeast genome. Trends Biotechnol 16(9):373–378

Osorio JS, Ji P, Drackley JK, Luchini D, Loor JJ (2013a) Supplemental Smartamine M or MetaSmart during the transition period benefits postpartal cow performance and blood neutrophil function. J Dairy Sci 96(10):6248–6263

Osorio JS, Trevisi E, Ballou MA, Bertoni G, Drackley JK, Loor JJ (2013b) Effect of the level of maternal energy intake prepartum on immunometabolic markers, polymorphonuclear leukocyte function, and neutrophil gene network expression in neonatal Holstein heifer calves. J Dairy Sci 96(6):3573–3587

Osorio JS, Ji P, Drackley JK, Luchini D, Loor JJ (2014a) Smartamine M and MetaSmart supplementation during the peripartal period alter hepatic expression of gene networks in 1-carbon metabolism, inflammation, oxidative stress, and the growth hormone-insulin-like growth factor 1 axis pathways. J Dairy Sci 97(12):7451–7464

Osorio JS, Trevisi E, Ji P, Drackley JK, Luchini D, Bertoni G, Loor JJ (2014b) Biomarkers of inflammation, metabolism, and oxidative stress in blood, liver, and milk reveal a better immunometabolic status in peripartal cows supplemented with Smartamine M or MetaSmart. J Dairy Sci 97(12):7437–7450

Pezeshki A, Muench GP, Chelikani PK (2012) Short communication: expression of peptide YY, proglucagon, neuropeptide Y receptor Y2, and glucagon-like peptide-1 receptor in bovine peripheral tissues. J Dairy Sci 95(9):5089–5094

Plata-Salaman CR (1998) Cytokines and feeding. News Physiol Sci 13:298–304
Plata-Salaman CR (2001) Cytokines and feeding. Int J Obes Relat Metab Disord 25(Suppl 5):S48–S52
Raboisson D, Caubet C, Tasca C, De Marchi L, Ferraton JM, Gannac S, Millet A, Enjalbert F, Schelcher F, Foucras G (2014) Effect of acute and chronic excesses of dietary nitrogen on blood neutrophil functions in cattle. J Dairy Sci 97(12):7575–7585
Reverchon M, Rame C, Cognie J, Briant E, Elis S, Guillaume D, Dupont J (2014) Resistin in dairy cows: plasma concentrations during early lactation, expression and potential role in adipose tissue. PLoS One 9(3), e93198
Reynolds CK, Aikman PC, Lupoli B, Humphries DJ, Beever DE (2003) Splanchnic metabolism of dairy cows during the transition from late gestation through early lactation. J Dairy Sci 86(4):1201–1217
Rico JE, Bandaru VV, Dorskind JM, Haughey NJ, McFadden JW (2015) Plasma ceramides are elevated in overweight Holstein dairy cows experiencing greater lipolysis and insulin resistance during the transition from late pregnancy to early lactation. J Dairy Sci 98(11):7757–7770
Rinaldi M, Ceciliani F, Lecchi C, Moroni P, Bannerman DD (2008) Differential effects of alpha1-acid glycoprotein on bovine neutrophil respiratory burst activity and IL-8 production. Vet Immunol Immunopathol 126(3-4):199–210
Rocco SM, McNamara JP (2013) Regulation of bovine adipose tissue metabolism during lactation. 7. Metabolism and gene expression as a function of genetic merit and dietary energy intake. J Dairy Sci 96(5):3108–3119
Roche JR, Lee JM, Macdonald KA, Berry DP (2007) Relationships among body condition score, body weight, and milk production variables in pasture-based dairy cows. J Dairy Sci 90(8):3802–3815
Roche JR, Friggens NC, Kay JK, Fisher MW, Stafford KJ, Berry DP (2009) Invited review: body condition score and its association with dairy cow productivity, health, and welfare. J Dairy Sci 92(12):5769–5801
Roche JR, Bell AW, Overton TR, Loor JJ (2013a) Nutritional management of the transition cow in the 21st century – a paradigm shift in thinking. Anim Prod Sci 53:1000–1023
Roche JR, Kay JK, Friggens NC, Loor JJ, Berry DP (2013b) Assessing and managing body condition score for the prevention of metabolic disease in dairy cows. The Veterinary clinics of North America. Food Anim Pract 29(2):323–336
Roche JR, Macdonald KA, Schutz KE, Matthews LR, Verkerk GA, Meier S, Loor JJ, Rogers AR, McGowan J, Morgan SR, Taukiri S, Webster JR (2013c) Calving body condition score affects indicators of health in grazing dairy cows. J Dairy Sci 96(9):5811–5825
Rukkwamsuk T, Wensing T, Geelen MJ (1999) Effect of overfeeding during the dry period on the rate of esterification in adipose tissue of dairy cows during the periparturient period. J Dairy Sci 82(6):1164–1169
Russell JB, Rychlik JL (2001) Factors that alter rumen microbial ecology. Science 292(5519): 1119–1122
Sander AK, Piechotta M, Schlamberger G, Bollwein H, Kaske M, Sipka A, Schuberth HJ (2011) Ex vivo phagocytic overall performance of neutrophilic granulocytes and the relation to plasma insulin-like growth factor-I concentrations in dairy cows during the transition period. J Dairy Sci 94(4):1762–1771
Saremi B, Mielenz M, Rahman MM, Hosseini A, Kopp C, Danicke S, Ceciliani F, Sauerwein H (2013) Hepatic and extrahepatic expression of serum amyloid A3 during lactation in dairy cows. J Dairy Sci 96(11):6944–6954
Sasaki K, Yamagishi N, Kizaki K, Sasaki K, Devkota B, Hashizume K (2014) Microarray-based gene expression profiling of peripheral blood mononuclear cells in dairy cows with experimental hypocalcemia and milk fever. J Dairy Sci 97(1):247–258
Schena M, Shalon D, Davis RW, Brown PO (1995) Quantitative monitoring of gene expression patterns with a complementary DNA microarray. Science 270(5235):467–470

Seo J, Osorio JS, Loor JJ (2013) Purinergic signaling gene network expression in bovine polymorphonuclear neutrophils during the peripartal period. J Dairy Sci 96(12):7675–7683

Shahzad K, Bionaz M, Trevisi E, Bertoni G, Rodriguez-Zas SL, Loor JJ (2014) Integrative analyses of hepatic differentially expressed genes and blood biomarkers during the peripartal period between dairy cows overfed or restricted-fed energy prepartum. PLoS One 9(6), e99757

Shonka BN, Tao S, Dahl GE, Spurlock DM (2015) Genetic regulation of prepartum dry matter intake in Holstein cows. J Dairy Sci 98(11):8195–8200

Soliman M, Kimura K, Ahmed M, Yamaji D, Matsushita Y, Okamatsu-Ogura Y, Makondo K, Saito M (2007) Inverse regulation of leptin mRNA expression by short- and long-chain fatty acids in cultured bovine adipocytes. Domest Anim Endocrinol 33(4):400–409

Sordillo LM (2016) Nutritional strategies to optimize dairy cattle immunity. J Dairy Sci 99(6):4967–4982

Sordillo LM, Babiuk LA (1991) Modulation of bovine mammary neutrophil function during the periparturient period following in vitro exposure to recombinant bovine interferon gamma. Vet Immunol Immunopathol 27(4):393–402

Sordillo LM, Raphael W (2013) Significance of metabolic stress, lipid mobilization, and inflammation on transition cow disorders. Vet Clin North Am Food Anim Pract 29(2):267–278

Sordillo LM, Contreras GA, Aitken SL (2009) Metabolic factors affecting the inflammatory response of periparturient dairy cows. Anim Health Res Rev 10(1):53–63

Spurlock DM, Dekkers JC, Fernando R, Koltes DA, Wolc A (2012) Genetic parameters for energy balance, feed efficiency, and related traits in Holstein cattle. J Dairy Sci 95(9):5393–5402

Steele MA, Schiestel C, AlZahal O, Dionissopoulos L, Laarman AH, Matthews JC, McBride BW (2015) The periparturient period is associated with structural and transcriptomic adaptations of rumen papillae in dairy cattle. J Dairy Sci 98(4):2583–2595

Sumner-Thomson JM, et al (2011) Differential expression of genes in adipose tissue of first-lactation dairy cattle. J Dairy Sci 94(1):361–369

Tetens J, Thaller G, Krattenmacher N (2014) Genetic and genomic dissection of dry matter intake at different lactation stages in primiparous Holstein cows. J Dairy Sci 97(1):520–531

Trevisi E, Amadori M, Cogrossi S, Razzuoli E, Bertoni G (2012) Metabolic stress and inflammatory response in high-yielding, periparturient dairy cows. Res Vet Sci 93(2):695–704

Vailati Riboni M, Kanwal M, Bulgari O, Meier S, Priest NV, Burke CR, Kay JK, McDougall S, Mitchell MD, Walker CG, Crookenden M, Heiser A, Roche JR, Loor JJ (2016) Body condition score and plane of nutrition prepartum affect adipose tissue transcriptome regulators of metabolism and inflammation in grazing dairy cows during the transition period. J Dairy Sci 99(1):758–770

Vailati-Riboni M, Kanwal M, Bulgari O, Meier S, Priest NV, Burke CR, Kay JK, McDougall S, Mitchell MD, Walker CG, Crookenden M, Heiser A, Roche JR, Loor JJ (2016) Body condition score and plane of nutrition prepartum affect adipose tissue transcriptome regulators of metabolism and inflammation in grazing dairy cows during the transition period. J Dairy Sci 99(1):758–770

Voelkerding KV, Dames SA, Durtschi JD (2009) Next-generation sequencing: from basic research to diagnostics. Clin Chem 55(4):641–658

Wang Y, Puntenney SB, Burton JL, Forsberg NE (2007) Ability of a commercial feed additive to modulate expression of innate immunity in sheep immunosuppressed with dexamethasone. Animal 1(7):945–951

Wang YQ, Puntenney SB, Burton JL, Forsberg NE (2009) Use of gene profiling to evaluate the effects of a feed additive on immune function in periparturient dairy cattle. J Anim Physiol Anim Nutr (Berl) 93(1):66–75

Wang X, Li X, Zhao C, Hu P, Chen H, Liu Z, Liu G, Wang Z (2012) Correlation between composition of the bacterial community and concentration of volatile fatty acids in the rumen during the transition period and ketosis in dairy cows. Appl Environ Microbiol 78(7):2386–2392

Wasinger VC, Cordwell SJ, Cerpa-Poljak A, Yan JX, Gooley AA, Wilkins MR, Duncan MW, Harris R, Williams KL, Humphery-Smith I (1995) Progress with gene-product mapping of the mollicutes: mycoplasma genitalium. Electrophoresis 16(7):1090–1094

Weber M, Locher L, Huber K, Rehage J, Tienken R, Meyer U, Danicke S, Webb L, Sauerwein H, Mielenz M (2016) Longitudinal changes in adipose tissue of dairy cows from late pregnancy to lactation. Part 2: The SIRT-PPARGC1A axis and its relationship with the adiponectin system. J Dairy Sci 99(2):1560–1570

Weber PS, et al (2001) Pre-translational regulation of neutrophil L-selectin in glucocorticoid-challenged cattle. Vet Immunol Immunopathol 83(3–4):213–240

Winter G, Kromer JO (2013) Fluxomics – connecting 'omics analysis and phenotypes. Environ Microbiol 15(7):1901–1916

Wong S, Pinkney J (2004) Role of cytokines in regulating feeding behaviour. Curr Drug Targets 5(3):251–263

Yuan K, Vargas-Rodriguez CF, Mamedova LK, Muckey MB, Vaughn MA, Burnett DD, Gonzalez JM, Titgemeyer EC, Griswold KE, Bradford BJ (2014) Effects of supplemental chromium propionate and rumen-protected amino acids on nutrient metabolism, neutrophil activation, and adipocyte size in dairy cows during peak lactation. J Dairy Sci 97(6):3822–3831

Zebeli Q, Ghareeb K, Humer E, Metzler-Zebeli BU, Besenfelder U (2015) Nutrition, rumen health and inflammation in the transition period and their role on overall health and fertility in dairy cows. Res Vet Sci 103:126–136

Zhang A, Sun H, Wang P, Han Y, Wang X (2012) Modern analytical techniques in metabolomics analysis. Analyst 137(2):293–300

Zhou Z, Bu DP, Vailati Riboni M, Khan MJ, Graugnard DE, Luo J, Cardoso FC, Loor JJ (2015) Prepartal dietary energy level affects peripartal bovine blood neutrophil metabolic, antioxidant, and inflammatory gene expression. J Dairy Sci 98(8):5492–5505

Systems Biology and Stem Cell Pluripotency: Revisiting the Discovery of Induced Pluripotent Stem Cell

Kaveh Mashayekhi, Vanessa Hall, Kristine Freude, Miya K Hoeffding, Luminita Labusca, and Poul Hyttel

Abstract

Recent breakthroughs in stem cell biology have accelerated research in the area of regenerative medicine. Over the past years, it has become possible to derive patient-specific stem cells which can be used to generate different cell populations for potential cell therapy. Systems biological modeling of stem cell pluripotency and differentiation have largely been based on prior knowledge of signaling pathways, gene regulatory networks, and epigenetic factors. However, there is a great need to extend the complexity of the modeling and to integrate different types of data, which would further improve systems biology and its uses in the field. In this chapter, we first give a general background on stem cell biology and regenerative medicine. Stem cell potency is introduced together with the hierarchy of stem cells ranging from pluripotent embryonic stem cells (ESCs) and induced pluripotent stem cells (iPSCs) to tissue-specific multipotent and unipotent stem cells. Secondly, we address some of the systems biological approaches which have

K. Mashayekhi
iGenomix S.L., Parc Cientific Universitat de Valencia,
Catedratico Agustin Escardino 9, Paterna 46980, Valencia, Spain

V. Hall • K. Freude • P. Hyttel (✉)
Department of Veterinary Clinical and Animal Sciences, University of Copenhagen,
Groennegaardsvej 7, Frederiksberg C DK-1870, Denmark
e-mail: poh@sund.ku.dk

M.K. Hoeffding
Copenhagen Consortium for Designer Organisms, University of Copenhagen,
Blegdamsvej 3B, Koebenhavn N DK-2200, Denmark

L. Labusca
Orthopedic and Traumatology Clinic, Emergency University Hospital, Saint Spiridon Piața Independenței 1, Iași 700111, Romania

National Institute of Research and Development for Technical Physics,
47 Mangeron Blvd., P.O. Box 833, P.O. 3, Iasi 700050, Romania

H.N. Kadarmideen (ed.), *Systems Biology in Animal Production and Health, Vol. 2*,
DOI 10.1007/978-3-319-43332-5_6

already added valuable knowledge to the stem cell field. Particular attention is paid to the most commonly used knowledge-based models as well as to the unsupervised data-driven model. Finally, we will revisit the discovery of the iPSCs by Yamanaka in 2006 and superimpose a data-driven systems biological approach on the data which this amazing discovery was based on. This approach helps to demonstrate how systems biology can complement the field of stem cell biology.

1 Introduction

Recent breakthroughs in stem cell biology have advanced the area of regenerative medicine and brought it closer to realization. Over the past years, it has become possible to derive patient-specific stem cells which can be used to generate different cell populations for in vitro cell modeling and potential cell therapy (Fig. 1). This development emphasizes the need to precisely understand and control stem cell pluripotency and subsequent differentiation of the pluripotent cell populations into a variety of target cell types. Differentiation of patient-specific stem cells can also be used for the establishment of patient-specific in vitro disease models, allowing for detailed molecular investigations of, for example, neurodegenerative diseases

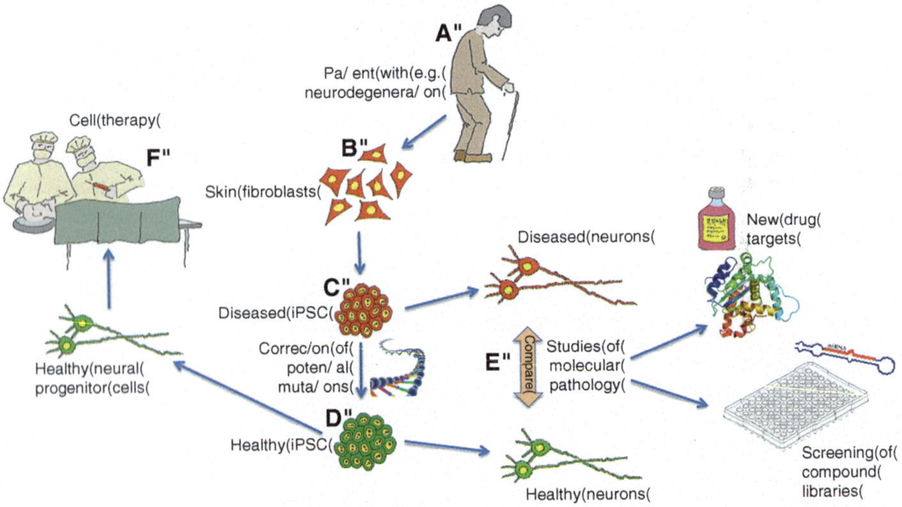

Fig. 1 Overview of the potential of human-induced pluripotent stem cells (*iPSC*). (**a**) Patient suffering from, e.g., neurodegenerative disorder like Parkinson's disease. (**b**) Fibroblasts cultured from skin biopsy. (c) iPSC reprogrammed from skin fibroblasts. (**d**) In the case of disorders caused by specific mutations, these may be corrected by CRISPR/Cas9 gene editing. (**e**) Healthy and diseased target cells, i.e., dopaminergic neurons in the case of, e.g., Parkinson's disease, can be compared and used as models for identifying new drug targets and for screening of compound libraries with potential effects on these targets. (**f**) As a future potential transplantation of therapeutic cell populations, i.e., dopaminergic neural precursors in the case of Parkinson's disease, is envisioned

affecting the target cells, i.e., the neurons. This was previously not possible and research was primarily based on animal models or cancer cell lines.

Pluripotency signaling and stem cell differentiation is governed by the activity of gene regulatory networks and tightly controlled by a multitude of pathways as well as by complex epigenetic mechanisms (Pir and Le Novere 2016). In the exponentially developing area of stem cell biology, technologies that deliver high-throughput and accurate molecular data result in the accumulation of large amounts of biological information. This data needs to be efficiently and meaningfully analyzed. Therefore, it is foreseen that the interface between stem cell biology and systems biology will continue to evolve. The extrapolation of more data from these cells may help to further refine in vitro protocols that will help drive stem cells towards the clinic. Models of stem cell pluripotency and differentiation have been developed for many years and have mostly been based on prior knowledge on the signaling pathways, gene regulatory networks, and epigenetic factors involved. However, there is a great need to extend the complexity of the modeling and to integrate different types of data, which would further improve systems biology and its uses in the field. This is where systems biology becomes an inevitable tool for further development of the field.

The first part of the chapter introduces the field of stem cell biology and regenerative medicine. The second part addresses some of the systems biological approaches which have already added valuable knowledge to the field of stem cell research. Finally, we revisit the discovery of the so-called induced pluripotent stem cells (iPSCs) by Yamanaka in 2006 (Takahashi and Yamanaka 2006) and superimpose a systems biological approach on the data in order to visualize how systems biology may streamline and accelerate methods in the field of stem cell biology.

2 Stem Cells

Stem cells are characterized by their ability to self-renew indefinitely, and by their ability to differentiate into tissue-specific cells when exposed to in vitro signaling and mechanical cues. A symmetrical division of a stem cell produces two identical daughter stem cells, whereas an asymmetrical division results in one stem cell and one cell which is destined for differentiation. Discrete stem cell populations exist in specialized niches of the body such as the bone marrow, the muscles, the gut, the skin, and the brain, where they can replace cells which have been lost due to injury, disease, or due to regular wear and tear. This ability to produce, maintain, and repair tissues and organs makes stem cells a very powerful tool in regenerative medicine and a subject of intense research all over the world (Singh et al. 2015).

2.1 Cell Potency in Early Human Embryonic Development

When the oocyte is released from the ovary and swept into the fallopian tube, it may be fertilized by a spermatozoon to generate the diploid zygote (Fig. 2). As the zygote travels towards the uterus, it divides and starts to compact, forming a ball of

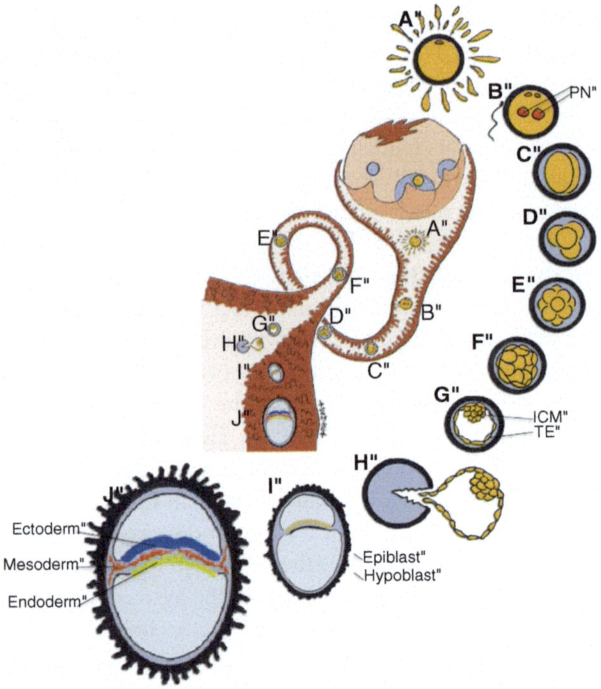

Fig. 2 Overview of human fertilization and initial embryonic development. (**a**) Ovulated mature oocyte. (**b**) One-celled fertilized ovum, i.e., zygote, with two haploid pronuclei (*PN*). (**c**) 2-cell stage. (**d**) 4-cell stage. (**e**) 8- to 16-cell stage. (**f**) Compacted morula where the outer cell layer forms a smooth surface of the embryo. (**g**) Blastocyst with cells divided into the first lineages, i.e., the inner cell mass (*ICM*) and trophectoderm (*TE*). (**h**) Blastocyst in the process of hatching from the zona pellucida. (**i**) Implanted blastocyst with the development of the epiblast and hypoblast from the inner cell mass. (**j**) Implanted embryo during gastrulation where the epiblast gives rise to the three germ layers: ectoderm, mesoderm, and endoderm

cells called the morula. Approximately five days after fertilization, a fluid-filled cavity forms within the morula which is now called the blastocyst. The blastocyst contains the pluripotent ICM, which becomes localized to one pole of the developing blastocyst, and an outer layer of trophectoderm. When the blastocyst has entered the uterus, it hatches from the zona pellucida, which previously encapsulated the oocyte, and implants into the uterine wall. The ICM then divides into two cell layers: the hypoblast and the pluripotent epiblast. At 14–16 days postfertilization, the epiblast undergoes gastrulation to generate the three germ layers: ectoderm, mesoderm, and endoderm. Each of these germ layers gives rise to multipotent cell populations such as neural progenitor cells from the ectoderm, hematopoietic stem cells from the mesoderm, and gut epithelial cells from the endoderm. In addition to the three germ layers, the epiblast also gives rise to the primordial germ cells which are the precursors of the germ cells in the developing embryo, i.e., the oocytes and the spermatogonia.

2.2 Cell Potency and Differentiation

Stem cell biology and developmental biology are intimately linked. In the paradigm of developmental biology, cells are divided into distinct categories according to their differentiation capacity, or "potency". In its strictest sense, a totipotent cell is one that can produce an entire fertile adult organism, including its temporary fetal membranes, when placed in the correct environment, such as the uterus, in the case of humans (De Paepe et al. 2014). The zygote is totipotent as are the daughter cells which are initially produced by symmetrical divisions of the zygote (Condic 2014). Due to their unlimited differentiation capacity, totipotent cells are often considered to be at the top of the developmental hierarchy. However, in the case of mammals, totipotency is not a cell state which can be sustained. During embryonic development, cells will only remain totipotent during the first few divisions of the zygote. Furthermore, no single mammalian cell can orchestrate the complex spatial and temporal patterns of proliferation, differentiation, and self-organization which result in an entire organism and its fetal membranes, when cultured in vitro (Condic 2014).

Pluripotency represents, in relation to totipotency, the next stage, which is reached once asymmetrical cell divisions of the totipotent cells result in the first differentiation and emergence of the blastocyst. Cells from the ICM are pluripotent; these can differentiate into all cell types of the body, but they do not contribute to the extraembryonic lineages derived from the trophectoderm under normal circumstances (Morgani et al. 2013). Embryonic stem cells (ESCs) are derived from the ICM, or the subsequent epiblast, of preimplantation embryos, and they can be cultured as self-renewing pluripotent cells in vitro, in principle, indefinitely (Morgani et al. 2013). Multipotent stem cells are generated in the embryo when pluripotent stem cells differentiate into specialized progenitor cells; examples of such are the neural progenitor cells which give rise to neurons and glia, or hematopoietic stem cells which give rise to different types of blood cells. These multipotent stem cells are lineage restricted, meaning they are committed to produce only certain cell types. Thus, they are more specialized than pluripotent stem cells and they can be found as "adult stem cells" in various organs and tissues postembryonic development (Roy and Kundu 2014). The most restricted stem cell types are referred to as unipotent, indicating that they only give rise to a single terminally differentiated cell type, such as a skin stem cell residing in the basal layer of the epidermis which only gives rise to keratinocytes.

2.3 Embryonic Stem Cells (ESCs)

ESCs were first derived from the ICM or the early epiblast of preimplantation mouse blastocysts in 1981 (Evans and Kaufman 1981; Martin 1981). When cultured in vitro on mitomycin-C inactivated fibroblasts, these mouse ESCs formed well-defined dome-shaped colonies composed of tightly packed small, round cells with a large nuclei and scanty cytoplasm. The ESCs were characterized by a high proliferative capacity and the ability to generate teratomas when injected into syngeneic mice, reminiscent of embryonic carcinoma cells (ECs). Additionally, when cultured in feeder-free conditions in vitro, the ESCs formed three-dimensional spheres called embryoid bodies which upon directed differentiation could develop into derivatives of all three germ layers. However, in contrast to ECs, the ESCs have a normal karyotype (Evans and Kaufman 1981; Martin 1981). Some years later, it was shown that ESCs could contribute to germ line chimeras (Bradley et al. 1984) and that aggregation of the ESCs together with developmentally compromised tetraploid embryos could give rise to fetuses exclusively composed of ESC-derived cells (Nagy et al. 1990). This unequivocally confirmed the pluripotent state of the cells. Like the ICM and the preimplantation epiblast of mice, the ESCs expressed pluripotent-related transcription factors including, *Oct4*, *Nanog*, *Sox2*, *Klf4* as well as *Rex1* and *Fgf4*, and ESCs derived from female mice were characterized by the presence of two active X chromosomes (Nichols and Smith 2009). It was discovered that ESCs could be cultured in feeder- and serum-free conditions by addition of the myeloid leukaemia inhibitory factor (LIF) and the bone morphogenic protein 4 (BMP4) to the culture medium (Williams et al. 1988; Ying et al. 2003). Additionally, it was found that the pluripotent state of murine ESCs could be promoted by blocking the MAPK/ERK pathway and by inhibiting GSK3-beta, both of which were known to induce differentiation of the ESCs (Nichols and Smith 2009; Yeo and Ng 2013). Accordingly, a LIF + 2i medium was developed, which allows for much better control and maintenance of the pluripotent ESCs in vitro (Ying et al. 2008).

Pluripotent stem cells have also been derived from postimplantation mouse epiblasts. In contrast to ESCs, these so-called epiblast stem cells (EpiSCs) are sustained in medium containing fibroblast growth factor (FGF) and the transforming growth factor (TGF)-beta family member Activin, but not LIF (Nichols and Smith 2009; Tesar et al. 2007). They depend on *Oct4*, *Sox2*, and *Nanog* expression and form embryoid bodies and teratomas, but display epigenetic silencing of one of the X chromosomes in female cells and cannot contribute to blastocyst chimaeras (Tesar et al. 2007). Additionally, EpiSCs actively transcribe lineage-specific genes, such as *Eomes*, *Gata6*, *Sox17*, and *T* (Brachyury), and do not express genes associated with the ICM, such as *Tbx3* and *Pecam1* (Tesar et al. 2007). In addition, EpiSCs form monolayers of columnar epithelium in culture, reminiscent of the rodent postimplantation epiblast (Nichols and Smith 2009). Thus, EpiSCs represent a later stage of development compared to ESCs, yet are still considered pluripotent. Accordingly, EpiSCs are classified as "primed" pluripotent stem cells in contrast to murine ESCs which are classified as "naïve" or belonging to the "ground state" of pluripotency (Nichols and Smith 2009).

Human ESCs were first derived from the ICM of preimplantation blastocysts in 1998 (Thomson et al. 1998). These cells were characterized by high levels of telomerase activity, indicative of an immortalized state, and could be sustained for more than 32 passages. They are able to differentiate into the three germ layers, when cultured as embryoid bodies, and give rise to teratomas, but for ethical reasons their ability to contribute to the germ line have not been tested. Human ESCs express surface markers such as stage-specific embryonic antigen (SSEA)-3 and (SSEA)-4, alkaline phosphatase, tumor rejection antigen (TRA)-1–60, and TRA-1–81, indicative of an undifferentiated state. However, later studies have shown that human ESCs are not equivalent to murine ESCs; rather, they display many features of the primed pluripotent state, similar to EpiSCs. For example, culturing of human ESCs requires TGF-beta/Activin signaling rather than LIF and BMP (Daheron et al. 2004; Xu et al. 2002; Xu et al. 2005; James et al. 2005), and human ESCs display a tendency to drive OCT4 expression via the proximal enhancer of this gene rather than the distal enhancer, which is the case in naïve ESCs (Hanna et al. 2010). Additionally, human ESCs share the flattened monolayer morphology of EpiSCs and most female human ESC lines display X chromosome inactivation (Hanna et al. 2010).

Regardless of these differences between mouse and human ESCs, both cell types are valuable research tools which can be used to study the mechanisms of cell fate decisions during development and disease. The pluripotent and proliferative capacity of ESCs makes it possible to generate large numbers of specific cell types in vitro, which would otherwise be difficult to obtain, such as cardiomyocytes, retinal pigment cells, and neurons (de Wert and Mummery 2003). Murine ESCs can be genetically modified and injected into blastocysts, thus giving rise to chimeras which upon mating will produce heterozygous and homozygous offspring. These offspring are useful for providing phenotypic readouts of the potential effects of specific mutations that can be introduced into the starting ESC populations (Capecchi 2005). Additionally, ESC-derived tissues can be applied for transplantation purposes. Indeed, it was reported in 2012 that retinal pigment epithelium derived from human ESCs is currently being investigated in early phases of several clinical trials for the treatment of macular degeneration (Trounson and McDonald 2015). Despite these advances, the use of human ESCs in research is a controversial subject (de Wert and Mummery 2003). The fact that these cells are derived from human preimplantation embryos, which have the potential to give life to a human being, poses an ethical dilemma for stem cell researchers. On the one side, the use of human ESCs can potentially alleviate suffering by providing knowledge about disease mechanisms and treatments. On the other side, the embryo must be destroyed in order to obtain the ESCs (de Wert and Mummery 2003). The later discovery of iPSCs provided researchers with a solution to this problem.

2.4 Reversal of Lineage Specification

In 1957, Conrad Waddington produced an "epigenetic landscape" modeling lineage specification during normal embryonic development (Waddington 1957). In this landscape, a marble residing on the top of a hill represented the totipotent or pluripotent cell. As the cell initiated the process of differentiation, it would slide down the hill in increasingly restricted paths, eventually terminating its movement at the lowest point of the landscape, representing its final differentiated state. Just as marbles are not likely to spontaneously roll back up to the top of a hill, it was believed that cells would not return to more potent states once terminally differentiated. Thus, lineage specification was considered unidirectional and irreversible in nature (Waddington 1957).

However, during the 1950s and 1960s, development of the somatic cell nuclear transfer (SCNT) technique demonstrated that the differentiated state of ectodermal, endodermal, and mesodermal amphibian cells could in fact be reset back to totipotency by introducing the nucleus of these cells into frog oocytes (Gurdon et al. 1958). The oocyte was found to contain molecules which were able to rewind the epigenetic landscape. A few decades later, Ian Wilmut and colleagues demonstrated that this principle could also be applied to mammals by producing Dolly the sheep, cloned by SCNT from a mammary gland epithelial cell (Wilmut et al. 1997). Together, these pioneering studies revealed that certain conditions in the oocyte could reverse cell differentiation and lineage specification. Thus, in the right environment, the marble could be induced to roll back up to the top of the hill.

2.5 Induced Pluripotent Stem Cells (iPSCs)

In a quest to unravel which factors in the oocyte were responsible for the dedifferentiation of somatic cells, Kazutoshi Takahashi and Shinya Yamanaka screened 24 candidate genes known to be important for embryonic stem cell identity (Takahashi and Yamanaka 2006). We will return to this experimental approach in the final section of this chapter. By retroviral transduction of different combinations of these genes into mouse embryonic fibroblasts (MEFs) they revealed that expression of only four transcription factors, *Oct4*, *Sox2*, *C-Myc*, and *Klf4*, was enough to convert the fibroblasts to pluripotent stem cells, when cultured in appropriate stem cell conditions (Takahashi and Yamanaka 2006). Soon thereafter, they were able to generate "induced" pluripotent stem cells (iPSCs) from human fibroblasts using the same four factors which are now commonly referred to as the Yamanaka factors (Takahashi et al. 2007). The ability to reverse the lineage specification of adult somatic cells back to the pluripotent state without the requirement for oocytes revolutionized the field of regenerative medicine and Yamanaka was consequently awarded with the Nobel Prize in Physiology or Medicine in 2012 alongside John Gurdon who undertook the previously mentioned frog SCNT experiments. Not only was it possible to avoid the ethical challenge associated with the isolation of ESCs from human blastocysts, it was now possible to obtain patient-specific pluripotent stem cells and differentiate these into any cell type of interest. In this way, inaccessible tissue, such as that of the brain, could now be studied in a dish and be applied for development of personalized medicine (Medvedev et al. 2010; Singh et al. 2015). Additionally, derivation of various tissues from patient-specific iPSCs allows for autologous cell transplantation for the treatment of various degenerative diseases and which circumvents immune rejection. Finally, reprogramming of differentiated cells into iPSCs provided a fascinating tool for researchers to study the basic mechanisms of cell fate conversions, both "forwards" and "backwards". Indeed, many groups have strived to map the paths which lead to the iPSC state and to identify techniques, as well as factors, which can push differentiated somatic cells back to the top of Waddington's epigenetic hill.

2.6 Reprogramming to iPSC: Techniques, Factors, and Cell Types

The desire to generate clinical grade iPSCs for transplantation purposes soon caused stem cell researchers to look for alternatives to the original retroviral reprogramming protocol used by Yamanaka's team. The retroviral delivery technique involved random integration of the Yamanaka transgenes into the host cell genome, causing a risk of insertional mutagenesis and potential problems with continuous expression or reactivation of the transgenes in differentiated iPSCs (Medvedev et al. 2010). In order to overcome these safety issues, a variety of reprogramming techniques were developed which involved only transient expression of the Yamanaka factors in the cells undergoing reprogramming. These techniques included the use of non-integrating vectors such as Adenovirus or Sendai virus, plasmids, DNA minicircles, PiggyBac transposons, episomal vectors or direct introduction of mRNA or protein into the cells. Although these alternative techniques were generally less efficient than reprogramming with retrovirus or lentivirus, they brought iPSC research one step closer to clinical application. In addition, several transcription factors were identified, which could replace one or more of the Yamanaka factors or enhance the reprogramming efficiency in the presence of the Yamanaka factors. For example, it was reported that *KLF4* and *C-MYC* could be substituted by the less oncogenic *NANOG* and *LIN28* in reprogramming of human cells (Yu et al. 2007), whereas the estrogen-related receptor *Esrrb* could substitute for *Klf4* when reprogramming MEFs (Feng et al. 2009). Furthermore, it was recently shown that members of the GATA family of transcription factors could substitute for *Oct4* in mouse cells, whereas *Sox2* could be replaced by the DNA replication inhibitor *Gmnn* (Shu et al. 2013). Also, a short hairpin suppression of p53 was able to enhance reprogramming efficiency without causing increased chromosomal instability (Rasmussen et al. 2014). Additionally, a myriad of chemical compounds have been shown to enhance reprogramming to iPSC. These include HDAC inhibitors such as valproic acid, sodium butyrate, and trichostatin A, TGF-beta inhibitors such as A-83-01 and SB43152, as well as inhibitors of MEK (PD0325901) and GSK3 (CHIR99021) (reviewed in Medvedev et al. 2010; Malik and Rao 2013).

Finally, researchers have broadened the repertoire of reprogrammable cell types to include not only fibroblasts, but also keratinocytes, peripheral blood T cells, hematopoietic stem cells, umbilical cord blood cells, renal epithelial cells, mesenchymal stem cells, mesenchymal stromal cells, hepatocytes, pancreatic islet beta cells, synovial cells, and other cell types (reviewed in Raab et al. 2014). These different cell types each present distinct advantages and disadvantages to reprogramming, according to their ease of derivation and maintenance in culture, and according to their proliferative capacity and endogenous expression of iPSC-related genes (Raab et al. 2014). Despite this, fibroblasts remain the most popular choice of cells for iPSC reprogramming.

There has been very great focus on the translation of the iPSC technology into therapeutic use. In 2015, the first clinical trials were initiated in Japan, where iPSC-derived retinal pigment epithelial cells were used for cell replacement therapy for age-related macular degeneration (Sheridan 2014). However, this was suspended only some months after initiation due to the discovery that the iPSCs contained chromosome abnormalities (webpage: https://www.ipscell.com/2015/07/firstipscstop/).

3 Systems Biology Approaches for Unraveling Stem Cell Pluripotency and Differentiation

Over the past years, different system biological approaches have been utilized for understanding and unraveling stem cell pluripotency and differentiation. Basically, two different types of modeling have been applied: knowledge-based models and data-driven models. Both of these models are based on well-documented data, but differ fundamentally in their handling of the data. The knowledge-based models are based on theories formulated from careful surveys of existing knowledge, whereas data-driven models are based on holistic unsupervised computational analyses of databases. In addition to these two types of models, completely theoretical models of cell fate and behavior may be applied in the field. In this chapter, the knowledge-based and data-driven models are discussed in detail.

3.1 Knowledge-Based Models Applied on Cell Pluripotency and Differentiation

The advantage of applying systems biology in stem cell biology has already materialized. Most of the available mechanistic models for stem cell biology and differentiation are built based on previous knowledge, also commonly described as the bottom-up approach. In this approach, the data to be included as well as their relationships are obtained from wet laboratory experiments, scientific literature, or public databases that contribute with previously generated modules of information that can be incorporated as building blocks into the modeling process (Le Novere 2015; Pir and Le Novere 2016). Knowledge-based modeling in stem cell biology has most extensively been applied in four areas: pluripotency-related signaling pathways, gene regulatory networks related to master pluripotency transcription factors, reprogramming of somatic cells into iPSCs, and epigenetic regulation. Each of these four aspects will be addressed in the following.

The first mathematical model applied to stem cell biology revealed that murine ESC self-renewal was dependent on the concentration of cytokines (Viswanathan et al. 2002). This was revealed not long time after the discovery that the LIF/JAK/STAT3 pathway was important for the maintenance of pluripotency (Niwa et al. 1998). This computational model generated predictions for the degree of self-renewal as a function of cytokine concentrations. These model predictions were consistent with experimental data and indicated that differences in the effects of LIF and another cytokine, hyper-interleukin-6 (HIL-6), were based on differences in receptor-binding stoichiometry and properties. These results revealed that ligand/receptor signaling thresholds could be used to model ESC fate.

The modeling approach of Viswanathan et al. (2002) described above was useful for examining cell fate decisions following a single cell division and under steady-state conditions. However, it did not permit direct application to the behavior at the more complex cell population dynamic level. To address this, the same research group extended the model substantially by incorporating cytokine-mediated regulation of single-cell proliferation, differentiation, and death to account for the clonal evolution of individual cells in a population. They also added cell-intrinsic parameters, such as the half-life of downstream transcription targets. The resulting model included heterogeneities in individual-cell properties (e.g., ligand–receptor complex numbers, cell cycle asynchrony, and transcription factor half-lives) to predict the generation of various progeny trajectories in response to diverse stimuli. Specifically, it became possible to integrate stochastic (individual cell) and deterministic (population-averaged) variables to compute dynamic cellular system behavior simultaneously at the individual cell as well as at subpopulation levels. The group employed a transgenic murine ESC line expressing GFP (green fluorescent protein) driven by the *Oct4* promoter, allowing for in situ tracking of pluripotency and differentiation. Using this cell line model, predictions were tested and were consistent with bulk measurements of *Oct4-GFP+* and *Oct4-GFP–* cell outputs over a range of exogenous conditions. This computational model provided alternatives to previous theories supporting the view that stem cell self-renewal versus differentiation choices are completely stochastic. Overall, the model indicates that the probability of self-renewal is neither a random (as predicted by the stochastic models) nor an invariant (as predicted by the Poisson model) property of individual cells, but more a consequence of the cell's dynamic interactions with its microenvironment. Finally, the model could be generalized to the previously characterized intestinal crypt system in elucidating relative contributions of stem and progenitor cells to population output.

Another modeling example is the deterministic model of self-renewing and differentiating ESC populations developed to predict the response to the cytokines LIF and FGF4 in addition to the extracellular matrix components laminin and fibronectin (Prudhomme et al. 2004b). Stem cell self-renewal versus differentiation fate decisions are difficult to characterize and analyze due to multiple competing rate processes occurring simultaneously among heterogeneous cell subpopulations. A mathematical model was described that allows flow cytometric measurement of population distributions of molecular markers to be deconvoluted. This, therefore, could address the cell population dynamics in terms of subpopulation-specific rate parameters and distinguish between commitment to differentiation, proliferation of differentiated cells, or proliferation of undifferentiated cells (i.e., self-renewal). This model was validated by means of dedicated, independent cell-tracking studies and demonstrated that it was capable of accurately interpreting relationships underlying the effects of external cues on cell responses in differentiating cultures via intracellular signals.

Woolf et al. (2005) adapted a Bayesian network learning algorithm to model proteomic signaling data for ESC fate responses to external cues (Woolf et al. 2005). They were able to characterize the signaling pathway influences as quantitative, logic-circuit type interactions. Their experimental data set included measurements for phosphorylation states of 28 signaling proteins across 16 different factorial combinations of cytokine and matrix stimuli. The group showed that Bayesian networks are able to organize data at two levels of abstraction. At the first level, the Bayesian network itself is a directed graph that reflects many known, physiological connections in the original source data. On a second level of abstraction, the connections between nodes can be plotted and these plots reveal novel insight into the underlying biochemical mechanisms. The approach also demonstrated that it could be possible to make reliable predictions of new conditions by performing experiments in silico in order to identify combinations of input parameters likely to yield interesting and useful results in the wet laboratory.

As described earlier, STAT3 signaling is involved in ESC self-renewal. Mahdavi et al. (2007) developed and validated a computational model of STAT3 pathway kinetics (Mahdavi et al. 2007). They revealed novel pathway responses such as overexpression of the receptor glycoprotein-130 results in reduced pathway activation and increased ESC differentiation. A systematic in silico screen was used to identify novel targets and protein interactions involved in STAT3 signaling and it was found that signaling activation and desensitization (the inability to respond to ligand re-stimulation) is regulated by balancing the activation state of a distributed set of parameters including nuclear export of STAT3, nuclear phosphatase activity, inhibition by suppressor of cytokine signaling, and receptor trafficking. This knowledge was used to devise a temporally modulated ligand delivery strategy that maximizes signaling activation and leads to enhanced ESC self-renewal.

Further investigations of the STAT3 effects on murine ESCs were performed by Moledina et al. (2012) using a combined in silico and experimental approach in which they directly manipulated, using laminar fluid flow, the local impact of endogenously secreted gp130-activating ligands and their activation of STAT3 signaling (Moledina et al. 2012). The model analysis predicted that flow-dependent changes in autocrine and paracrine ligand binding would impact heterogeneity in cell- and colony-level STAT3 signaling activation and cause a gradient of cell fate determination along the direction of flow. Interestingly, analysis also predicted that local cell density would be inversely proportional to the degree to which endogenous secretion contributed to cell fate determination. Experimental validation using functional activation of STAT3 by secreted factors under microfluidic perfusion culture demonstrated that STAT3 activation, and consequently ESC fate, could in fact be manipulated by flow rate, position in the flow field, and local cell organization.

Peerani et al. (2009) extended a stochastic model developed previously to predict the fraction of autocrine and paracrine trajectories captured by a single cell in cell culture assays in order to study how STAT3 activation is modulated by three ESC culture parameters: colony size, colony separation, and degree of clustering. The results of this modeling and associated wet laboratory experiments demonstrated that colonies less than 100 μm in diameter were too small to maximize endogenous STAT3 activation and that colonies separated by more than 400 μm could be considered independent from each other (Peerani et al. 2009). This resulted in defined parameter boundaries for the use of ESCs in screening studies and demonstrated the importance of context in stem cell responsiveness to exogenous cues. It also revealed that niche size is an important parameter in stem cell fate control.

Ellison et al. (2009) developed a computational model assuming a critical need for cell-secreted survival factors to better characterize possible effects of autocrine and paracrine signaling in murine ESCs (Ellison et al. 2009). This model suggested a precise way in which the removal of putative survival factors could affect stem cell survival in culture. The predictions were experimentally tested in murine ESCs by taking advantage of a novel microfluidic device allowing removal of the cell-conditioned medium at defined time intervals. Experimental results in both serum-containing and defined media confirmed the computational model predictions, suggesting the existence of unknown survival factors with distinct rates of diffusion, and revealed an adaptive/selective phase in the response of the ESCs to a lack of paracrine signaling.

Yeo et al. (2013) developed a multiscale mathematical model describing population-segregated growth kinetics, metabolism, and the expression of selected pluripotency genes to characterize nutritional requirements for murine ESCs (Yeo et al. 2013). The model was validated by wet laboratory experiments with the expansion of undifferentiated murine ESCs encapsulated in hydrogels in batch and perfusion cultures using bioreactors. The model clearly demonstrated that despite sufficient nutrient and growth factor provision, the accumulation of inhibitory metabolites resulted in the unscheduled differentiation of ESCs and a decline in their cell numbers in the batch cultures. In contrast, perfusion cultures maintained metabolite concentration below toxic levels, resulting in the robust expansion (>16-fold) of high-quality "naïve" ESCs within four days.

Gene regulatory networks are extremely important in cell fate determination during embryonic development as well as in stem cell biology. Transcription factors are the major players in the regulation of gene expression and very often bind to each other's promoters, establishing gene regulatory feed-back loops (Pir and Le Novere 2016). Computational models have also addressed these networks and master pluripotency transcription factors.

Krupinski et al. (2011) developed a computational modeling framework for mimicking murine blastocyst formation. They concluded that the coupling of gene expression with the mechanics of cell movement is important for formation of both the trophectoderm and the endoderm (Krupinski et al. 2011). Further, Bessonnard et al. (2014) developed a model that describes the temporal dynamics of ERK signaling and of the concentrations of NANOG, GATA6, secreted FGF4 and FGF receptor 2 in the differentiation of the ICM into the epiblast and primitive endoderm. The model reveals a mechanism relying on the co-existence between three stable steady states (tristability), which correspond to ICM, epiblast, and primitive endoderm (Bessonnard et al. 2014).

The OCT4–SOX2–NANOG network is of major importance in murine ESCs. Chikarmane et al. (2006) designed a kinetic modeling approach that ascribes function to this network by making plausible assumptions about the interactions between the transcription factors at the gene promoter binding sites and RNA polymerase, at each of the three genes as well as at the target genes (Chickarmane et al. 2006).

They identified a bistable switch in the network which arises due to several positive feedback loops and is switched on/off by input environmental signals. The switch stabilizes the expression levels of the three genes and through their regulatory roles on the downstream target genes leads to a binary decision: when OCT4, SOX2, and NANOG are expressed and the switch is on, the self-renewal genes are on and the differentiation genes are off. The opposite holds when the switch is off. The model was subsequently further extended to include more transcription factors (Chickarmane and Peterson 2008).

Murine ESC populations are heterogeneous with respect to the expression of NANOG and Glauche et al. (2010) applied a novel mathematical transcription factor network model to explore mechanisms and feedback regulations to describe the effect of this differential expression (Glauche et al. 2010). They were able to show that these variations can occur under different assumptions yielding similar experimental characteristics. Based on model predictions, experimental strategies were designed to distinguish between these explanations. The authors concluded that the heterogeneity with respect to the NANOG expression is most likely a functional element to control the differentiation propensity of an ESC population. Furthermore, a conceptual framework that consistently explains NANOG variability and a potential "gatekeeper" function of NANOG expression with respect to the control of ESC differentiation was proposed. The conclusions were later supported by other modeling approaches (Chickarmane et al. 2012). Subsequent modeling approaches have demonstrated that autocrine FGF feedback can establish distinct states of NANOG expression murine ESCs and may be an underlying background for these effects (Lakatos et al. 2014). Other reports on murine ESC modeling also conclude that interaction between NANOG expression and FGF4/Erk signaling qualifies as a key mechanism to manipulate ESC pluripotency (Herberg et al. 2014).

Other modeling studies help to show the complexity of cell signaling pathways which interact with OCT4, which is a master regulator of pluripotency. Munoz et al. (2013) combined single-cell quantitative immunofluorescence microscopy and gene expression analysis together with theoretical modeling (Munoz Descalzo et al. 2013). They found that a network of protein complexes, including among others NANOG, OCT4, TCF3, and β-catenin, are important for maintaining ESC pluripotency. The results suggest that the function of the network is to buffer the transcriptional activity of OCT4 under different conditions.

Our current understanding of how somatic cells are reprogrammed into iPSCs has improved over recent years, although certain details remain incomplete. There is a general understanding that reprogramming is either a two- or three-step process (Samavarchi-Tehrani et al. 2010; Golipour et al. 2012; Hansson et al. 2012) that involves an early stochastic phase, whereby the cell increases proliferation, undergoes metabolic changes, initiates the mesenchymal-to-epithelial transition, changes its expression of histone marks, and activates both DNA repair and RNA processing. Later events are marked by maturation and stabilization phases, whereby activation of the core pluripotency circuit is initiated, among several other cellular changes (Buganim et al. 2013). Despite this recent knowledge, a rate-limiting step in iPSC reprogramming exists. That is, only a few cells of the starting population are successfully reprogrammed and reasons for this remain unknown. There are currently two conflicting theories on the reprogramming mechanisms which infer that reprogrammed cells arise from either a selective cell type in the cell population or emerge equally from all cell types present in the starting population. In the former case, the so-called elite model proposes that iPSCs can be exclusively generated from a subpopulation of cells with particular reprogramming competences (Byrne et al. 2009; Wakao et al. 2011), while in the latter case, the stochastic model proposes that every cell type has the potential to be reprogrammed (Yamanaka 2009). The stochastic model for iPSC reprogramming was supported by a computational model of transcriptional control of cell fate specification (MacArthur et al. 2008). The model comprises two mutually interacting subcircuits: A central pluripotency circuit consisting of interactions between stem cell-specific transcription factors OCT4, SOX2, and NANOG coupled to a differentiation circuit consisting of interactions between lineage-specifying master genes. The model suggests that under certain circumstances, amplification of low-level fluctuations in transcriptional status (transcriptional "noise") may be sufficient to trigger reactivation of the core pluripotency switch and reprogramming to a pluripotent state. Further quantitative analyses have defined distinct cell-division-rate-dependent and -independent modes for accelerating the stochastic course of reprogramming and suggest that the number of cell divisions is a key parameter driving epigenetic reprogramming to pluripotency (Hanna et al. 2009).

Clinical applications of human iPSCs are critically dependent on efficient upscaling of cells required for differentiation into relevant therapeutic cell populations. The process of upscaling of human iPSCs generation has also been the focus of ordinary differential equation (ODE) based modeling by Selekman et al. (2013). These authors demonstrate a strategy for investigating the efficiency and scalability of iPSC differentiation platforms. Using two previously reported epithelial differentiation systems, they fitted an ODE-based kinetic model to data representing dynamics of various cell subpopulations in the culture by estimating rate constants of each cell subpopulation's cell fate decisions (self-renewal, differentiation, death). Sensitivity analyses on predicted rate constants indicated which cell fate decisions had the greatest impact on overall epithelial cell yield in each differentiation process. In this way, the group outlined a novel approach for quantitative analysis of established laboratory-scale human iPSC differentiation systems, which may ease development to produce large quantities of cells for tissue engineering applications (Selekman et al. 2013).

As described previously in this chapter, Waddington proposed epigenetic regulatory mechanisms already in 1957 and they have been shown to be of utmost importance in relation to cell differentiation and reprogramming to pluripotency. Epigenetics is defined as relatively stable and potentially heritable changes in the cell phenotype without any changes in the DNA sequence. At the molecular level, epigenetic mechanisms work through small molecules such as methyl-groups deposited on DNA, the 3D chromatin structure dictated by numerous modifications of a set of DNA-binding proteins such as histones, together with small molecules deposited on the DNA-binding proteins, and non-coding RNA with regulatory functions (Boland et al. 2014). In particular, the reprogramming of somatic cells into iPSCs has been a focus for systems biological modeling of epigenetic mechanisms, as it was discovered that epigenetic factors might be a major barrier in the reprogramming of somatic cells into iPSCs (Papp and Plath 2011).

A systems biological approach for modeling iPSC reprogramming which also takes into account epigenetic mechanisms has been presented by Artyomov et al. (2010). Here, a cell cycle-based binary model including both gene regulatory networks and permissive as well as repressive epigenetic marks was constructed. The results obtained from the model are consistent with the stochastic mode of iPSC reprogramming and identified the rare pathways that allow reprogramming to occur. If validated by further experiments, this model could be developed further for designing optimal strategies for reprogramming (Artyomov et al. 2010). The model developed by Artyomov et al. (2010) was further refined by Hu et al. (2011) who proposed a novel Markov model in which they calculated the reprogramming rate and showed that it would increase in the condition of knockdown of somatic transcription factors or inhibition of global DNA methylation (Hu et al. 2011). The utility of this latter model was verified by testing it with the real dynamic gene expression data spanning across different intermediate stages in the iPSC reprogramming process.

An extension of these studies was performed by Flottmann et al. (2012) who pushed probabilistic Boolean network approaches further by focusing on the interplay between gene expression, chromatin modifications, and DNA methylation. The simulation results showed good reproduction of experimental observations during reprogramming and indicated that faster changes in DNA methylation increased the speed of reprogramming at the expense of efficiency, while accelerated chromatin modifications moderately improved efficiency (Flottmann et al. 2012).

Interestingly, modeling that also includes epigenetic data has also added to the discussion of whether iPSC reprogramming occurs according to the elite model or the stochastic model described earlier in this chapter. In particular, Grácio et al. (2013) built mass-action models of the core regulatory elements controlling iPSC reprogramming, which included not only the network of transcription factors NANOG, OCT4, and SOX2 but also important epigenetic regulatory features of DNA methylation and histone modification (Gracio et al. 2013). This work suggested an alternative, somehow intermediate hypothesis that the unpredictability and variation in reprogramming decreases as the cell progresses along the induction process and that identifiable groups of cells with elite-seeming behavior can emerge from a stochastic process.

Systems biological modeling including the epigenetic effects has also contributed to the understanding of the variability in NANOG expression, which was described earlier in this chapter. Sasai et al. (2013) constructed a model of the core gene network of mouse ESCs and showed that the phenotypic heterogeneity of ESCs can be explained by a slow transcriptional switching of the chromatin permissiveness at the *Nanog* locus related to the cell cycle progression (Sasai et al. 2013). These NANOG-related changes simulated ESCs to fluctuate among multiple transient states and triggered differentiation into lineage-specific cell states. The epigenetic landscape underlying these transitions was calculated and it was proposed that the slow *Nanog* switching was the underlying mechanism for the change in ESC states. This proposal was further supported by Zhang and Wolynes (2014) who constructed an approximation that allows for quantitative modeling of the epigenetic network (Zhang and Wolynes 2014).

Altogether, the knowledge-based systems biological modeling has contributed significantly to our understanding of the cell pluripotency and differentiation and has supported existing theories but also led to the generation of new theories on how iPSC reprogramming occurs (for a review, see (Muraro et al. 2013)).

3.2 Data-Driven Models Applied on Cell Pluripotency and Differentiation

The models presented in the previous section are knowledge-based and built on well-documented information about the systems to be investigated. Constructing such knowledge-based models can be tedious, as the relevant molecular interactions have to be extracted from wet laboratory experiments or from the literature. In an emerging field like stem cell research, there is also the possibility that the list of interactions derived from the literature is incomplete (Pir and Le Novere 2016). The following section will focus on the less abundant examples of models built without prior information, i.e., built using data-driven holistic top-down approaches to identify network components or network structures related to stem cell biology where no previous hypotheses have been formulated. With the tremendous and exponentially increasing speed by which high-throughput data is accumulating, such data-driven unsupervised methods are becoming more and more attractive in systems biology, as they do not require tedious wet laboratory experimentation and literature surveys, but instead rely on holistic, non-biased analysis of the accumulating data mass.

To gain insight into murine ESC differentiation, Prudhomme et al. (2004a) conducted a data-driven multivariate proteomic analysis of murine ESC self-renewal versus differentiation signaling. Phosphorylation states of 31 intracellular signaling network proteins were obtained from 16 conditions by quantitative Western blotting. Partial least squares modeling was then applied to determine which components were most strongly correlated with cell proliferation and differentiation constants obtained following flow cytometry of OCT4 expression. This approach yielded, in a data-driven manner, a set of seven phospho-proteins (STAT3, RAF1, MEK, ERK, SRC, PKCε, and PKBα) most critically associated with cell differentiation rates and/or proliferation rates. Many of the predictions were found to be consistent with the previous literature or experimental tests that were later performed (Prudhomme et al. 2004b).

Data-driven approaches have also been applied on gene regulatory networks. Using generalized singular value decomposition and comparative partition around medoids algorithms, a set of transcription factors, including FOX, GATA, MYB, NANOG, OCT, PAX, SOX, and STAT, and the FGF response element were identified as key regulators underlying the transcriptional co-expression maintaining pluripotency (Sun et al. 2008). By transcriptional intervention conducted in silico, dynamic behavior of pathways was examined, demonstrating how much and in which specific ways each gene or gene combination affected the behavior transition of a pathway in response to ESC differentiation or pluripotency induction.

Chavez et al. (2009) further performed a data-driven in silico identification of a core regulatory network of OCT4 in human ESCs using an approach where they carried out an integrated analysis of high-throughput data (ChIP-on-chip and RNAi experiments along with promoter sequence analysis of putative target genes) and identified a core OCT4 regulatory network in human ESCs consisting of 33 target genes (Chavez et al. 2009). Likewise, Gu et al. (2008) performed in silico analyses of novel miRNA candidates and miRNA–mRNA pairs in ESCs and were able to identify a plethora of previously unknown miRNA candidates, including 545 RNAs that are enriched in ESCs compared with adult cells (Gu et al. 2008).

Qin et al. (2014) employed a data-driven systematic model to elucidate endogenous barriers limiting this process of human iPSC reprogramming (Qin et al. 2014). They systematically dissected cellular reprogramming by combining a genomewide RNAi screen, innovative computational methods, extensive single-hit validation, and mechanistic investigation of relevant pathways and networks. They succeeded in identifying reprogramming barriers, including genes involved in transcription, chromatin regulation, ubiquitination, dephosphorylation, vesicular transport, and cell adhesion. Specifically, disintegrin and metalloproteinase (ADAM) proteins inhibit reprogramming together with clathrin-mediated endocytosis, which positively regulates TGF-β signaling.

4 Revisiting Yamanaka's Discovery of Induced Pluripotent Stem Cells (iPSCs) by a Data-Driven Top-Down Systems Biological Approach

Generation of induced pluripotent stem cells (iPSCs) in the mouse was first reported in 2006 by Takahashi and Yamanaka, which revolutionized the field of stem cell biology. The process that Yamanaka used to reveal the four key transcription factors required to produce iPSCs was an eloquent and systematic approach. Yamanaka selected in total 24 candidate genes (Table 1). The selection of these genes was based on previous publications which showed that they played some role in maintaining pluripotency in mouse and human ESCs. Yamanaka tested the effect of the 24 genes on inducing pluripotency in Fbx15$^{ßgeo/ßgeo}$ mouse fibroblasts. These transgenic mice contained a fused form of the ß-galactosidase and neomycin resistance genes, located in the *Fbx15* gene. This meant that only reprogrammed cells positively expressing *Fbx15* (a specific gene expressed in mouse ESCs and early embryos) would survive when cultured in the presence of G418 antibiotic. This was a particularly elegant strategy used to select for only pluripotent cell colonies. Yamanaka's group used retrovirus to transduce single genes or combinations of the genes into fibroblasts. He discovered that transduction of the genes individually did not result in any ESC-like colonies; however, when all 24 genes were included, ESC-like colonies emerged. In order to determine which genes were of greatest importance, Yamanaka subsequently systematically withdrew individual genes from the pool of 24 to identify ten genes which were important for the ESC-like colony formation. Also, he found that the combination of these ten genes produced

Table 1 Yamanaka's starting list of 24 genes

Number of Genes	Gene Symbol	Mouse TF	Cell Lineage Identification	Hedgehog	P53	WNT Targets	WNT	Cell Cycle	VEGFs	Embryonic Stem Cells	Stem Cell Signalling	Homeobox	Stem Cell TF	EMT	Jak-STAT	Stem Cells	Terminal Diff Marker	Notch Sig	Epigen Chrom Remod Fact	Notch Targets	Total number of pathways
1	Myc*	•			+	+	+			+			+		+	+					7
2	Stat3	•								+	+		+	+	+						5
3	Sox2*	•	+			+				+			+			+					5
4	Oct4*	•	+			+				+		+	+								5
5	Ctnnb1	•		+			+				+			+				+			5
6	Nanog	•				+	+				+			+				+			5
7	Klf4*	•								+	+				+						3
8	Utf1	•								+											1
9	Sox15	•								+											1
10	Sall4	•								+											1
11	Gdf3		+							+											2
12	Dppa2									+											1
13	Dppa3									+											1
14	Dppa4									+											1
15	Dppa5a																				0
16	Gm14458																				0
17	Fthl17																				0
18	Grb2																				0
19	Tcl1																				0
20	Fbxo15																				0
21	Eras																				0
22	Dnmt3l																				0
23	ECAT1																				0
24	ECAT8																				0
Table 2	Pathway Population		3	1	1	4	3	0	0	12	4	1	4	3	3	2	0	2	0	0	[illegible]

Here we have ranked the genes according to our pathway analyses presented in Table 2, from ascending to descending based on the number of pathways an individual gene relates to. A + indicates that the gene is associated with the pluripotency-related pathways listed. Gray highlights the top seven genes based on their involvement in more than three pluripotent-related pathways. All four of the Yamanaka factors fall into the top seven and are marked in bold with *

more colonies than the transduction with all 24 genes. Again, Yamanaka performed withdrawal of individual genes from the pool of ten to identify four genes, i.e., *Oct4*, *Sox2*, *Klf4*, and *c-Myc*, which together could produce ESC-like colonies with a similar efficiency as the combination of the ten genes. Further subtraction of genes from these four (i.e., combinations of two or three) failed to give rise to ESC-like colonies. He thus concluded that *Oct4*, *Klf4*, *Sox2*, and *c-Myc* are key reprogramming factors for producing iPSCs cells from mouse fibroblasts. It is clear that Yamanaka and his colleagues invested an enormous effort in identifying the 24 genes, but even more in performing the reductive wet laboratory experiments which involved vector construction, transduction, and subsequent analyses at each step. The 24 genes selected appear to be rather arbitrarily chosen. How Yamanaka managed to discover these four magic factors seems to be rather incredulous. Of the several thousand genes expressed in pluripotent cells it was amazing that he could select only 24 and from these discover the four required to reprogram a somatic cell. In fact, it turns out that there are other genes that can also reprogram somatic cells and/or enhance the reprogramming process (e.g., *PDGF-BB*, *LIN28*, *LMYC*), and even other combinations of genes can be used (Jung et al. 2014; Park et al. 2014).

We were interested in examining whether a holistic data-driven systems biology approach could, as an alternative and precise method, be applied to large "omic" data sets available on pluripotent cell populations in order to discover the genes important for pluripotency and cell reprogramming. Hence, we asked the question: Can we, using systems biology, identify the key Yamanaka factors using a data-driven, holistic, unbiased approach or will we find another gene set? Consequently, in the following, we analyze a transcriptomic data set of cultured murine ESCs in order to identify crucial genes important for pluripotency by means of an unbiased holistic systems biology approach. We decided to perform this analysis only on transcriptomic data as a "proof of principle", but the approach should have optimally combined transcriptomic data with proteomic, lipidomic and other "omic" data sets. We evaluated two data sets of mouse ESCs representing two very different physiological conditions already addressed earlier in this chapter: Naïve murine ESCs are representative of the in vivo early epiblast derived from the blastocyst and are maintained in vitro in the presence of the growth factor leukemia inhibitory factor (LIF). These are thought to represent an early state of pluripotency termed the "ground state" of pluripotency (Nichols and Smith 2009). In contrast, primed murine ESCs, also often referred to as EpiSCs since they can be derived from the later stage epithelial epiblast, are cultured in the presence of bFGF (Nichols and Smith 2009). In the following systems biological analysis, we performed in silico analyses on a publicly available data set obtained from Gene Expression Omnibus (GEO). We selected RNA sequencing (RNA-seq) data on both naïve (mouse ESC) and primed ESCs (here referred to as EpiLCs) from the GEO entry GSE67259 and used four of the 74 samples, mESC #1 and #2 and EpiLC #1 and #2. This particular data set was recently published by Yamanaka and colleagues (Sasaki et al. 2015). The RNA-seq was performed on an ABI SOLiD 5500XL genetic analyzer. The reads were processed by trimming library adaptor and poly-A sequences by cut adapt-1.3 and trimmed reads of less than 30 bp were discarded. The remaining reads were mapped onto the genomes and were separated from the ERCC spike-in RNAs using a Perl script and the Cufflinks 2.2.0 program. Finally, the reads mapped on the ERCC spike-in RNAs were used to estimate transcript copy numbers per cell, and expression levels were normalized to the total mapped reads (Sasaki et al. 2015). We then further normalized the raw data and determined the differentially expressed genes comparing naïve vs. primed ESCs. From a total number of 26,311 transcripts, 2079 passed a filter of greater than or equal to 1.5-fold and a *P*-value of less than 0.05 using the Benjamini and Hochberg false discovery rate. From these 2079 genes, 1100 were upregulated and 979 downregulated in naïve compared to the primed state (Fig. 3).

Subsequently, we performed pathway enrichment on the genes that were upregulated in the naïve mouse ESCs. The selection criteria used included that a gene (1) had to be involved in at least three of the selected 19 stem cell and pluripotency enrichment pathways available (Pathway Central Qiagen) and (2) had to be a transcription factor. Pathway enrichment criterium 1 resulted in a narrowing of the genes of interest to a list of 40 and application of the criterium 2 further narrowed the list to 18 genes (Table 2). The data was further sorted by binary search sorting to create an ordered list from ascending to descending based on the number of

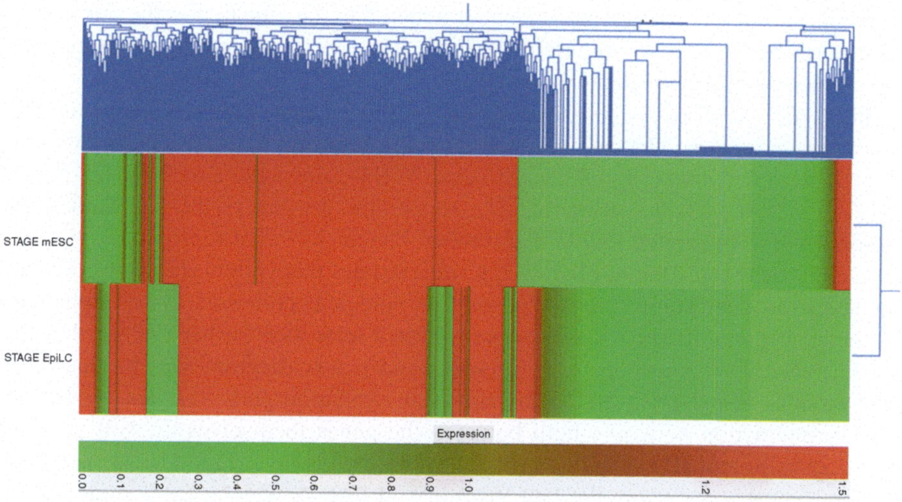

Fig. 3 Heat map of differentially expressed genes in mouse ESCs representing naïve pluripotency (*mESC*) versus primed pluripotency (*EpiLC*). The heat map shows hierarchical clustering of 2079 genes that passed Benjamini and Hochberg with a false discovery rate test correction ($p < 0.05$) and which had an expression alteration of greater than or equal to 1.5-fold

Table 2 A list of 18 genes were, by means of pathway analyses, selected from 1100 that were upregulated in naïve vs. primed state ESCs

Number of Genes	Gene Symbol	Fold Change	Mouse TF	Cell Lineage Identification	Hedgehog	P53	WNT Targets	WNT	Cell Cycle	VEGFs	Embryonic Stem Cells	Stem Cell Signalling	Homeobox	Stem Cell TF	EMT	Jak-STAT	Stem Cells	Terminal Diff Marker	Notch Sig	Epigen Chrom Remod Fact	Notch Targets	Total number of pathways
1	Myc*	0.3	•			+	+	+			+			+		+	+					7
2	Nanog	16.9	•				+	+				+			+				+			5
3	Stat3	6.2	•								+	+		+	+	+						5
4	Sox2*	5.8	•	+			+				+			+			+					5
5	Cdx2	25.0	•	+			+				+			+								4
6	Gli2	2.9	•	+			+									+					+	4
7	Smad1	1.9	•			+			+							+			+			4
8	Klf4*	504.5	•								+	+				+						3
9	Tbx3	1143.0	•		+							+		+								3
10	Otx2	0.2	•									+	+	+								3
11	Fzd2	0.2	•								+			+								2
12	Hoxc12	316.2	•										+	+								2
13	Rbl1	2.4	•						+			+										2
14	Ezh2	1.5	•											+						+		2
15	Hoxc4	25.0	•			+			+													2
16	Nr0b1	6.2	•											+								1
17	Hes1	3.0	•								+											1
18	Utf1	0.3	•								+											1
Table 1	Pathway Population			3	1	3	5	2	3	0	8	6	2	10	2	5	2	0	2	1	1	56

The selection criteria included that a gene (1) had to be involved in at least three of the selected 19 stem cell and pluripotency enrichment pathways available (Pathway Central Qiagen) and (2) had to be a transcription factor. The genes are ranked from ascending to descending based on the number of scores they have in different pluripotent-related cell pathways. A + indicates that the gene is associated with the pluripotency-related pathways listed. Grey highlights the top 10 genes based on their involvement in more than three pluripotent-related pathways. Three of the four Yamanaka factors fall into this list and are marked in bold with * different pluripotency pathways the genes adhered to, i.e., the higher the number of pathways, the higher the ranking of the gene. The top 10 genes fell into more than three pluripotent-related pathways.

Interestingly, 10 of the 18 genes we report fall into one pathway, the stem cell transcription factors (TF) pathway, followed by the embryonic stem cells pathway (8), the stem cell signaling pathway (6), the WNT pathway (5), and the Jak-STAT (5) (Table 2). It was interesting that *Myc* topped the list (one of the four Yamanaka factors). We were excited to find that three of Yamanaka's four reprogramming factors fell into the top 10 of our list, which substantiates that systems biology could be successful in determining master genes for biological processes to be applied in biomedical research. The top 10 genes in Table 2 were then further evaluated using the GeneNetworkCentralPro hub (Qiagen) and a flux diagram was produced in PathVisio to reveal predictive transcriptional regulation networks and predictive protein interaction networks both between the genes themselves as well as with other genes potentially related to the pluripotent pathways (Fig. 4). The network flux was constructed using the PathVisio in xml format, which can be migrated to other systems biology hubs such as Cytoscape for further interface to external web

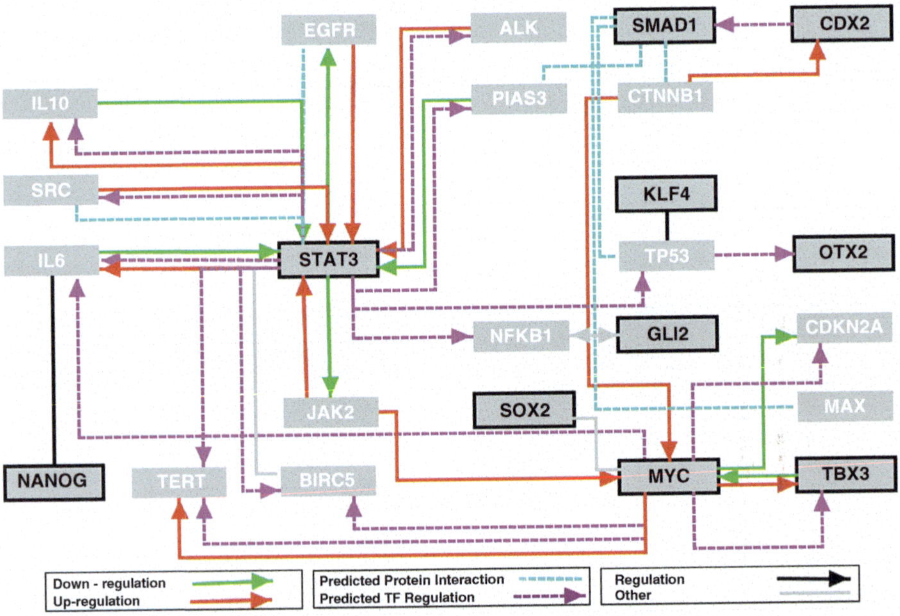

Fig. 4 Flux diagram of top 10 ranked genes from Table 2. The diagram reveals a broad insight into common gene interaction network and their flow based on the interaction data obtained from GNCPro, SABiosciences. The interactions can be either undirected (mutual) as gene–gene or protein–protein interactions or directed, where one gene or protein can alter the expression of its neighbor by upregulation/downregulation or acting as a transcription factor. Boxes outlined in black are the target genes and light gray boxes their immediate neighbors. The network flux is a construction by PathVisio in xml format for migration to other systems biology hubs

resources and other available public data repositories like GEO, ArrayExpress, UniProt, Reactume, Entrez, and Gene Ontology.

It is important to mention that one reprogramming gene, *Oct4*, did not fall into our final list of 18 genes. This is related to its lower *P*-value, which did not pass the minimum cutoff fold change (1.5) based on the set minimum criteria. It is also important to mention that the number of replicates in this experiment is also lower than that of the minimum number of replicates ($n=2$) for a non-parametric, non-Gaussian distribution which requires a minimum of five replicates. Thus, a higher number of replicates may have increased the *P*-value of this and potentially other genes. However, the mere fact that *Oct4* is known to be expressed in both naïve and primed mouse ESCs (Buecker et al. 2014) helps to substantiate the fact that this gene has not been selected by our analyses which were based on genes that were upregulated early on in pluripotency (i.e., in the naïve-like condition).

Additionally, it is important to note this approach utilized a selection for pluripotency transcription factors from the assumption that these master regulators of gene expression would be most likely to engage in the reprogramming process. This may or may not be the case. Several other genes are represented in the list of 18 genes predicted to be important for the reprogramming events. It is not known whether overexpression of these may actually reprogram somatic cells more efficiently than the Yamanaka factors. Therefore, it is obvious that this predictive tool must be used in combination with wet laboratory experiments to test whether combinations in the overexpression of these other genes can replace and/or improve the known reprogramming factors in effectively reprogramming somatic cells.

Finally, we decided to apply the same ranking method as we applied to our shortlist of 18 genes (Table 2) from the RNA-seq data onto Yamanaka's starting gene list ($n=24$) that he used for his wet laboratory experiments (Table 1). This resulted in seven of the 24 genes falling into more than three different pluripotent-related pathways (Table 1). In these top seven genes, all of the four Yamanaka factors (*Oct4*, *Sox2*, *Klf4*, and *Myc*) were found. Of greater interest was that two others, *Stat3* and *Nanog*, were also found in both the top seven of Yamanaka's list (Table 1) and in the top 10 of our gene list (Table 2). Although *Nanog* has been used to improve reprogramming, this has not been shown to be true for *Stat3*, which does not appear to improve reprogramming. Therefore, again, it is important to state that although these in silico tools are helpful, the genes that were brought forward require validations and biological testing to confirm the ability to reprogram somatic cells.

5 Conclusions and Further Challenges

Systems biological models based on computational approaches have already shown that they can elevate the field of stem cell biology. Knowledge-based bottom-up applications have shed significant light on mechanisms underlying stem cell pluripotency and differentiation. Results arising from unsupervised data-driven top-down approaches are fewer, but have contributed to the understanding of the gene regulatory networks underlying pluripotency. In order to fully benefit from the synergy between systems biology and stem cell research, there are some challenges to be overcome: First, common data repositories of a variety of "omics" data on pluripotency and differentiation in man and different animal species must be expanded. Secondly, faithful algorithms for unsupervised analyses must be refined and shared. And finally, a communication gap between systems biologists on the one hand and stem cell researchers on the other must continuously be narrowed through synergistic collaborative efforts.

Acknowledgments We thank the following for financial support: The Danish National Advanced Technology Foundation (project number 047-2011-1; patient-specific stem cell-derived models for Alzheimer's disease) and the European Union 7th Framework Program (PIAP-GA-2012-324451-STEMMAD) and Innovation Fund Denmark, BrainStem.

References

Artyomov MN, Meissner A, Chakraborty AK (2010) A model for genetic and epigenetic regulatory networks identifies rare pathways for transcription factor induced pluripotency. PLoS Comput Biol 6(5), e1000785. doi:10.1371/journal.pcbi.1000785

Bessonnard S, De Mot L, Gonze D, Barriol M, Dennis C, Goldbeter A, Dupont G, Chazaud C (2014) Gata6, Nanog and Erk signaling control cell fate in the inner cell mass through a tristable regulatory network. Development 141(19):3637–3648. doi:10.1242/dev.109678

Boland MJ, Nazor KL, Loring JF (2014) Epigenetic regulation of pluripotency and differentiation. Circ Res 115(2):311–324. doi:10.1161/CIRCRESAHA.115.301517

Bradley A, Evans M, Kaufman MH, Robertson E (1984) Formation of germ-line chimaeras from embryo-derived teratocarcinoma cell lines. Nature 309(5965):255–256

Buecker C, Srinivasan R, Wu Z, Calo E, Acampora D, Faial T, Simeone A, Tan M, Swigut T, Wysocka J (2014) Reorganization of enhancer patterns in transition from naive to primed pluripotency. Cell Stem Cell 14(6):838–853. doi:10.1016/j.stem.2014.04.003

Buganim Y, Faddah DA, Jaenisch R (2013) Mechanisms and models of somatic cell reprogramming. Nat Rev Genet 14(6):427–439. doi:10.1038/nrg3473

Byrne JA, Nguyen HN, Reijo Pera RA (2009) Enhanced generation of induced pluripotent stem cells from a subpopulation of human fibroblasts. PLoS One 4(9), e7118. doi:10.1371/journal.pone.0007118

Capecchi MR (2005) Gene targeting in mice: functional analysis of the mammalian genome for the twenty-first century. Nat Rev Genet 6(6):507–512. doi:10.1038/nrg1619

Chavez L, Bais AS, Vingron M, Lehrach H, Adjaye J, Herwig R (2009) In silico identification of a core regulatory network of OCT4 in human embryonic stem cells using an integrated approach. BMC Genomics 10:314. doi:10.1186/1471-2164-10-314

Chickarmane V, Peterson C (2008) A computational model for understanding stem cell, trophectoderm and endoderm lineage determination. PLoS One 3(10), e3478. doi:10.1371/journal.pone.0003478

Chickarmane V, Troein C, Nuber UA, Sauro HM, Peterson C (2006) Transcriptional dynamics of the embryonic stem cell switch. PLoS Comput Biol 2(9), e123. doi:10.1371/journal.pcbi.0020123

Chickarmane V, Olariu V, Peterson C (2012) Probing the role of stochasticity in a model of the embryonic stem cell: heterogeneous gene expression and reprogramming efficiency. BMC Syst Biol 6:98. doi:10.1186/1752-0509-6-98

Condic ML (2014) Totipotency: what it is and what it is not. Stem Cells Dev 23(8):796–812. doi:10.1089/scd.2013.0364

Daheron L, Opitz SL, Zaehres H, Lensch MW, Andrews PW, Itskovitz-Eldor J, Daley GQ (2004) LIF/STAT3 signaling fails to maintain self-renewal of human embryonic stem cells. Stem Cells 22(5):770–778. doi:10.1634/stemcells.22-5-770

De Paepe C, Krivega M, Cauffman G, Geens M, Van de Velde H (2014) Totipotency and lineage segregation in the human embryo. Mol Hum Reprod 20(7):599–618. doi:10.1093/molehr/gau027

de Wert G, Mummery C (2003) Human embryonic stem cells: research, ethics and policy. Hum Reprod 18(4):672–682

Ellison D, Munden A, Levchenko A (2009) Computational model and microfluidic platform for the investigation of paracrine and autocrine signaling in mouse embryonic stem cells. Mol Biosyst 5(9):1004–1012. doi:10.1039/b905602e

Evans MJ, Kaufman MH (1981) Establishment in culture of pluripotential cells from mouse embryos. Nature 292(5819):154–156

Feng B, Jiang J, Kraus P, Ng JH, Heng JC, Chan YS, Yaw LP, Zhang W, Loh YH, Han J, Vega VB, Cacheux-Rataboul V, Lim B, Lufkin T, Ng HH (2009) Reprogramming of fibroblasts into induced pluripotent stem cells with orphan nuclear receptor Esrrb. Nat Cell Biol 11(2):197–203. doi:10.1038/ncb1827

Flottmann M, Scharp T, Klipp E (2012) A stochastic model of epigenetic dynamics in somatic cell reprogramming. Front Physiol 3:216. doi:10.3389/fphys.2012.00216

Glauche I, Herberg M, Roeder I (2010) Nanog variability and pluripotency regulation of embryonic stem cells--insights from a mathematical model analysis. PLoS One 5(6), e11238. doi:10.1371/journal.pone.0011238

Golipour A, David L, Liu Y, Jayakumaran G, Hirsch CL, Trcka D, Wrana JL (2012) A late transition in somatic cell reprogramming requires regulators distinct from the pluripotency network. Cell Stem Cell 11(6):769–782. doi:10.1016/j.stem.2012.11.008

Gracio F, Cabral J, Tidor B (2013) Modeling stem cell induction processes. PLoS One 8(5), e60240. doi:10.1371/journal.pone.0060240

Gu P, Reid JG, Gao X, Shaw CA, Creighton C, Tran PL, Zhou X, Drabek RB, Steffen DL, Hoang DM, Weiss MK, Naghavi AO, El-daye J, Khan MF, Legge GB, Wheeler DA, Gibbs RA, Miller JN, Cooney AJ, Gunaratne PH (2008) Novel microRNA candidates and miRNA-mRNA pairs in embryonic stem (ES) cells. PLoS One 3(7), e2548. doi:10.1371/journal.pone.0002548

Gurdon JB, Elsdale TR, Fischberg M (1958) Sexually mature individuals of Xenopus laevis from the transplantation of single somatic nuclei. Nature 182(4627):64–65

Hanna J, Saha K, Pando B, van Zon J, Lengner CJ, Creyghton MP, van Oudenaarden A, Jaenisch R (2009) Direct cell reprogramming is a stochastic process amenable to acceleration. Nature 462(7273):595–601. doi:10.1038/nature08592

Hanna J, Cheng AW, Saha K, Kim J, Lengner CJ, Soldner F, Cassady JP, Muffat J, Carey BW, Jaenisch R (2010) Human embryonic stem cells with biological and epigenetic characteristics similar to those of mouse ESCs. Proc Natl Acad Sci U S A 107(20):9222–9227. doi:10.1073/pnas.1004584107

Hansson J, Rafiee MR, Reiland S, Polo JM, Gehring J, Okawa S, Huber W, Hochedlinger K, Krijgsveld J (2012) Highly coordinated proteome dynamics during reprogramming of somatic cells to pluripotency. Cell Rep 2(6):1579–1592. doi:10.1016/j.celrep.2012.10.014

Herberg M, Kalkan T, Glauche I, Smith A, Roeder I (2014) A model based analysis of culture-dependent phenotypes of mESCs. PLoS One 9(3), e92496. doi:10.1371/journal.pone.0092496

Hu Z, Qian M, Zhang MQ (2011) Novel Markov model of induced pluripotency predicts gene expression changes in reprogramming. BMC Syst Biol 5(Suppl 2):S8. doi:10.1186/1752-0509-5-S2-S8

James D, Levine AJ, Besser D, Hemmati-Brivanlou A (2005) TGFbeta/activin/nodal signaling is necessary for the maintenance of pluripotency in human embryonic stem cells. Development 132(6):1273–1282. doi:10.1242/dev.01706

Jung L, Tropel P, Moal Y, Teletin M, Jeandidier E, Gayon R, Himmelspach C, Bello F, Andre C, Tosch A, Mansouri A, Bruant-Rodier C, Bouille P, Viville S (2014) ONSL and OSKM cocktails act synergistically in reprogramming human somatic cells into induced pluripotent stem cells. Mol Hum Reprod 20(6):538–549. doi:10.1093/molehr/gau012

Krupinski P, Chickarmane V, Peterson C (2011) Simulating the mammalian blastocyst--molecular and mechanical interactions pattern the embryo. PLoS Comput Biol 7(5), e1001128. doi:10.1371/journal.pcbi.1001128

Lakatos D, Travis ED, Pierson KE, Vivian JL, Czirok A (2014) Autocrine FGF feedback can establish distinct states of Nanog expression in pluripotent stem cells: a computational analysis. BMC Syst Biol 8:112. doi:10.1186/s12918-014-0112-4

Le Novere N (2015) Quantitative and logic modelling of molecular and gene networks. Nat Rev Genet 16(3):146–158. doi:10.1038/nrg3885

MacArthur BD, Please CP, Oreffo RO (2008) Stochasticity and the molecular mechanisms of induced pluripotency. PLoS One 3(8), e3086. doi:10.1371/journal.pone.0003086

Mahdavi A, Davey RE, Bhola P, Yin T, Zandstra PW (2007) Sensitivity analysis of intracellular signaling pathway kinetics predicts targets for stem cell fate control. PLoS Comput Biol 3(7), e130. doi:10.1371/journal.pcbi.0030130

Malik N, Rao MS (2013) A review of the methods for human iPSC derivation. Methods Mol Biol 997:23–33. doi:10.1007/978-1-62703-348-0_3

Martin GR (1981) Isolation of a pluripotent cell line from early mouse embryos cultured in medium conditioned by teratocarcinoma stem cells. Proc Natl Acad Sci U S A 78(12): 7634–7638

Medvedev SP, Shevchenko AI, Zakian SM (2010) Induced pluripotent stem cells: problems and advantages when applying them in regenerative medicine. Acta Naturae 2(2):18–28

Moledina F, Clarke G, Oskooei A, Onishi K, Gunther A, Zandstra PW (2012) Predictive microfluidic control of regulatory ligand trajectories in individual pluripotent cells. Proc Natl Acad Sci U S A 109(9):3264–3269. doi:10.1073/pnas.1111478109

Morgani SM, Canham MA, Nichols J, Sharov AA, Migueles RP, Ko MS, Brickman JM (2013) Totipotent embryonic stem cells arise in ground-state culture conditions. Cell Rep 3(6):1945–1957. doi:10.1016/j.celrep.2013.04.034

Munoz Descalzo S, Rue P, Faunes F, Hayward P, Jakt LM, Balayo T, Garcia-Ojalvo J, Martinez Arias A (2013) A competitive protein interaction network buffers Oct4-mediated differentiation to promote pluripotency in embryonic stem cells. Mol Syst Biol 9:694. doi:10.1038/msb.2013.49

Muraro MJ, Kempe H, Verschure PJ (2013) Concise review: the dynamics of induced pluripotency and its behavior captured in gene network motifs. Stem Cells 31(5):838–848. doi:10.1002/stem.1340

Nagy A, Gocza E, Diaz EM, Prideaux VR, Ivanyi E, Markkula M, Rossant J (1990) Embryonic stem cells alone are able to support fetal development in the mouse. Development 110(3):815–821

Nichols J, Smith A (2009) Naive and primed pluripotent states. Cell Stem Cell 4(6):487–492. doi:10.1016/j.stem.2009.05.015

Niwa H, Burdon T, Chambers I, Smith A (1998) Self-renewal of pluripotent embryonic stem cells is mediated via activation of STAT3. Genes Dev 12(13):2048–2060

Papp B, Plath K (2011) Reprogramming to pluripotency: stepwise resetting of the epigenetic landscape. Cell Res 21(3):486–501. doi:10.1038/cr.2011.28

Park SJ, Yeo HC, Kang NY, Kim H, Lin J, Ha HH, Vendrell M, Lee JS, Chandran Y, Lee DY, Yun SW, Chang YT (2014) Mechanistic elements and critical factors of cellular reprogramming revealed by stepwise global gene expression analyses. Stem Cell Res 12(3):730–741. doi:10.1016/j.scr.2014.03.002

Peerani R, Onishi K, Mahdavi A, Kumacheva E, Zandstra PW (2009) Manipulation of signaling thresholds in "engineered stem cell niches" identifies design criteria for pluripotent stem cell screens. PLoS One 4(7), e6438. doi:10.1371/journal.pone.0006438

Pir P, Le Novere N (2016) Mathematical models of pluripotent stem cells: at the dawn of predictive regenerative medicine. Methods Mol Biol 1386:331–350. doi:10.1007/978-1-4939-3283-2_15

Prudhomme W, Daley GQ, Zandstra P, Lauffenburger DA (2004a) Multivariate proteomic analysis of murine embryonic stem cell self-renewal versus differentiation signaling. Proc Natl Acad Sci U S A 101(9):2900–2905. doi:10.1073/pnas.0308768101

Prudhomme WA, Duggar KH, Lauffenburger DA (2004b) Cell population dynamics model for deconvolution of murine embryonic stem cell self-renewal and differentiation responses to cytokines and extracellular matrix. Biotechnol Bioeng 88(3):264–272. doi:10.1002/bit.20244

Qin H, Diaz A, Blouin L, Lebbink RJ, Patena W, Tanbun P, LeProust EM, McManus MT, Song JS, Ramalho-Santos M (2014) Systematic identification of barriers to human iPSC generation. Cell 158(2):449–461. doi:10.1016/j.cell.2014.05.040

Raab S, Klingenstein M, Liebau S, Linta L (2014) A comparative view on human somatic cell sources for iPSC generation. Stem Cells Int 2014:768391. doi:10.1155/2014/768391

Rasmussen MA, Holst B, Tumer Z, Johnsen MG, Zhou S, Stummann TC, Hyttel P, Clausen C (2014) Transient p53 suppression increases reprogramming of human fibroblasts without affecting apoptosis and DNA damage. Stem Cell Rep 3(3):404–413. doi:10.1016/j.stemcr.2014.07.006

Roy S, Kundu TK (2014) Gene regulatory networks and epigenetic modifications in cell differentiation. IUBMB Life 66(2):100–109. doi:10.1002/iub.1249

Samavarchi-Tehrani P, Golipour A, David L, Sung HK, Beyer TA, Datti A, Woltjen K, Nagy A, Wrana JL (2010) Functional genomics reveals a BMP-driven mesenchymal-to-epithelial transition in the initiation of somatic cell reprogramming. Cell Stem Cell 7(1):64–77. doi:10.1016/j.stem.2010.04.015

Sasai M, Kawabata Y, Makishi K, Itoh K, Terada TP (2013) Time scales in epigenetic dynamics and phenotypic heterogeneity of embryonic stem cells. PLoS Comput Biol 9(12), e1003380. doi:10.1371/journal.pcbi.1003380

Sasaki K, Yokobayashi S, Nakamura T, Okamoto I, Yabuta Y, Kurimoto K, Ohta H, Moritoki Y, Iwatani C, Tsuchiya H, Nakamura S, Sekiguchi K, Sakuma T, Yamamoto T, Mori T, Woltjen K, Nakagawa M, Yamamoto T, Takahashi K, Yamanaka S, Saitou M (2015) Robust in vitro induction of human germ cell fate from pluripotent stem cells. Cell Stem Cell 17(2):178–194. doi:10.1016/j.stem.2015.06.014

Selekman JA, Das A, Grundl NJ, Palecek SP (2013) Improving efficiency of human pluripotent stem cell differentiation platforms using an integrated experimental and computational approach. Biotechnol Bioeng 110(11):3024–3037. doi:10.1002/bit.24968

Sheridan C (2014) Stem cell therapy clears first hurdle in AMD. Nat Biotechnol 32(12):1173–1174. doi:10.1038/nbt1214-1173

Shu J, Wu C, Wu Y, Li Z, Shao S, Zhao W, Tang X, Yang H, Shen L, Zuo X, Yang W, Shi Y, Chi X, Zhang H, Gao G, Shu Y, Yuan K, He W, Tang C, Zhao Y, Deng H (2013) Induction of pluripotency in mouse somatic cells with lineage specifiers. Cell 153(5):963–975. doi:10.1016/j.cell.2013.05.001

Singh VK, Kalsan M, Kumar N, Saini A, Chandra R (2015) Induced pluripotent stem cells: applications in regenerative medicine, disease modeling, and drug discovery. Front Cell Dev Biol 3:2. doi:10.3389/fcell.2015.00002

Sun Y, Li H, Liu Y, Mattson MP, Rao MS, Zhan M (2008) Evolutionarily conserved transcriptional co-expression guiding embryonic stem cell differentiation. PLoS One 3(10), e3406. doi:10.1371/journal.pone.0003406

Takahashi K, Yamanaka S (2006) Induction of pluripotent stem cells from mouse embryonic and adult fibroblast cultures by defined factors. Cell 126(4):663–676. doi:10.1016/j.cell.2006.07.024

Takahashi K, Tanabe K, Ohnuki M, Narita M, Ichisaka T, Tomoda K, Yamanaka S (2007) Induction of pluripotent stem cells from adult human fibroblasts by defined factors. Cell 131(5):861–872. doi:10.1016/j.cell.2007.11.019

Tesar PJ, Chenoweth JG, Brook FA, Davies TJ, Evans EP, Mack DL, Gardner RL, McKay RD (2007) New cell lines from mouse epiblast share defining features with human embryonic stem cells. Nature 448(7150):196–199. doi:10.1038/nature05972

Thomson JA, Itskovitz-Eldor J, Shapiro SS, Waknitz MA, Swiergiel JJ, Marshall VS, Jones JM (1998) Embryonic stem cell lines derived from human blastocysts. Science 282(5391):1145–1147

Trounson A, McDonald C (2015) Stem cell therapies in clinical trials: progress and challenges. Cell Stem Cell 17(1):11–22. doi:10.1016/j.stem.2015.06.007

Viswanathan S, Benatar T, Rose-John S, Lauffenburger DA, Zandstra PW (2002) Ligand/receptor signaling threshold (LIST) model accounts for gp130-mediated embryonic stem cell self-renewal responses to LIF and HIL-6. Stem Cells 20(2):119–138. doi:10.1634/stemcells.20-2-119

Waddington C (1957) The strategy of the genes, Routledge, NY
Wakao S, Kitada M, Kuroda Y, Shigemoto T, Matsuse D, Akashi H, Tanimura Y, Tsuchiyama K, Kikuchi T, Goda M, Nakahata T, Fujiyoshi Y, Dezawa M (2011) Multilineage-differentiating stress-enduring (Muse) cells are a primary source of induced pluripotent stem cells in human fibroblasts. Proc Natl Acad Sci U S A 108(24):9875–9880. doi:10.1073/pnas.1100816108
Williams RL, Hilton DJ, Pease S, Willson TA, Stewart CL, Gearing DP, Wagner EF, Metcalf D, Nicola NA, Gough NM (1988) Myeloid leukaemia inhibitory factor maintains the developmental potential of embryonic stem cells. Nature 336(6200):684–687. doi:10.1038/336684a0
Wilmut I, Schnieke AE, McWhir J, Kind AJ, Campbell KH (1997) Viable offspring derived from fetal and adult mammalian cells. Nature 385(6619):810–813. doi:10.1038/385810a0
Woolf PJ, Prudhomme W, Daheron L, Daley GQ, Lauffenburger DA (2005) Bayesian analysis of signaling networks governing embryonic stem cell fate decisions. Bioinformatics 21(6):741–753. doi:10.1093/bioinformatics/bti056
Xu RH, Chen X, Li DS, Li R, Addicks GC, Glennon C, Zwaka TP, Thomson JA (2002) BMP4 initiates human embryonic stem cell differentiation to trophoblast. Nat Biotechnol 20(12):1261–1264. doi:10.1038/nbt761
Xu RH, Peck RM, Li DS, Feng X, Ludwig T, Thomson JA (2005) Basic FGF and suppression of BMP signaling sustain undifferentiated proliferation of human ES cells. Nat Methods 2(3):185–190. doi:10.1038/nmeth744
Yamanaka S (2009) Elite and stochastic models for induced pluripotent stem cell generation. Nature 460(7251):49–52. doi:10.1038/nature08180
Yeo JC, Ng HH (2013) The transcriptional regulation of pluripotency. Cell Res 23(1):20–32. doi:10.1038/cr.2012.172
Yeo D, Kiparissides A, Cha JM, Aguilar-Gallardo C, Polak JM, Tsiridis E, Pistikopoulos EN, Mantalaris A (2013) Improving embryonic stem cell expansion through the combination of perfusion and Bioprocess model design. PLoS One 8(12), e81728. doi:10.1371/journal.pone.0081728
Ying QL, Nichols J, Chambers I, Smith A (2003) BMP induction of Id proteins suppresses differentiation and sustains embryonic stem cell self-renewal in collaboration with STAT3. Cell 115(3):281–292
Ying QL, Wray J, Nichols J, Batlle-Morera L, Doble B, Woodgett J, Cohen P, Smith A (2008) The ground state of embryonic stem cell self-renewal. Nature 453(7194):519–523. doi:10.1038/nature06968
Yu J, Vodyanik MA, Smuga-Otto K, Antosiewicz-Bourget J, Frane JL, Tian S, Nie J, Jonsdottir GA, Ruotti V, Stewart R, Slukvin II, Thomson JA (2007) Induced pluripotent stem cell lines derived from human somatic cells. Science 318(5858):1917–1920. doi:10.1126/science.1151526
Zhang B, Wolynes PG (2014) Stem cell differentiation as a many-body problem. Proc Natl Acad Sci U S A 111(28):10185–10190. doi:10.1073/pnas.1408561111

The manufacturer's authorised representative in the EU is Springer Nature Customer Service Centre GmbH, Europaplatz 3, 69115 Heidelberg, Germany. If you have any concerns regarding our products, please contact ProductSafety@springernature.com

Printed and bound by CPI Group (UK) Ltd, Croydon, CR0 4YY
05/10/2026
02228251-0001